DECORATIVE AND INNOVATIVE USE OF CONCRETE

Graham True
Ph.D, MICT

Whittles Publishing

CRC Press
Taylor & Francis Group

Published by
Whittles Publishing,
Dunbeath,
Caithness KW6 6EG,
Scotland, UK
www.whittlespublishing.com

Distributed in North America by
CRC Press LLC,
Taylor and Francis Group,
6000 Broken Sound Parkway NW, Suite 300,
Boca Raton, FL 33487, USA

© 2012 Graham True
ISBN 978-1904445-48-7

USA ISBN 978-1-4398-7643-5

Printed by Studio RBB, Latvia

CONTENTS

Decorative and Innovative use of Concrete

Concrete has become the most widely used construction material. However, the intrinsic nature of it allows the production of both 'good' and 'bad' concrete from the very same group of raw materials. It is the knowledge and experience (or lack) of those who use the material that often affects the quality of the concrete that we see around us.

So often the opportunity may be lost for not only providing a durable design in concrete but also in a form that is both pleasing to the eye and has the ability to weather suitably, and blend in or, if needed, to stand out and provide a needed contrast.

This book traces the use of concrete as a decorative and innovative material, utilising its unique properties. Designers from all over the world have been able to capitalise on the mouldability and final rigid stone-like form of concrete to furnish our environment with a variety of buildings and structures. This may range from thin but very strong structures to art forms that will initially appear as an implausible use of the material.

This book includes numerous processes of how concrete can be, or was formulated, as well as finishes to the final product, because these all provide us with an understanding as to what is possible with the material. Stains, applied colouring, deep abraded textures, transparency, movement, simulation and even spray applied metal coatings have all been mentioned, indicating the vast range of possibilities.

It is hoped that this book will enlighten those who wish to take the time to discover the potential residing within the nature of concrete, helping them to provide new and novel decorative uses and adaptations.

PREFACE

Concrete does not have to be a dull material –
any dullness comes from a lack of the users' and designers'
imagination, combined with an inability to see the
fundamental potential of the material. I hope this book
shows that concrete can do almost anything we want it to.

Graham True PhD., MICT.
GFT Materials Consultancy

Decorative and Innovative use of Concrete

The author wishes to express his appreciation to Whittles Publishing for considering and undertaking the publishing of this book, and to Mike Walker (retired Technical Director, The Concrete Society) for first tabling the concept of a book on decorative concrete and proposing the author for compiling the draft.

Considerable help in finding suitable reference illustrations has been provided by the staff of Whittles Publishing, in particular, Shelley Teasdale to whom I am particularly grateful.

I also wish to thank Edwin Trout, Information Services, The Concrete Society, for providing numerous illustrations and approval to use some of the historical process diagrams that accompany the text. It is through some of the now neglected practices that we can appreciate what can be undertaken using concrete.

The text and illustrations have attached acknowledgements, however, particular thanks is expressed to the following who have contributed outstanding examples of their work:

William (Bill) Mitchell. Many thanks for a most enjoyable day with you at your office in Harrods and for background and further illustrations of your work.

Christopher Stanley, who through Edwin Trout, agreed to the inclusion of illustrations from his book *Highlights in the History of Concrete*.

Carole Vincent, for her enthusiasm, comments and agreement to inclusion of her work.

Jannette Ireland, for superb illustrations through stages and final examples of her work.

Terry Pawson Associates, for providing and agreeing to the inclusion of specific examples of their work.

ACKNOWLEDGEMENTS

Lafarge Ductal, for excellent illustrations showing the potential of their development.

Litracon, for an excellent example of their light – transmitting concrete.

Beco Wallform, Quad-Lock and Polysteel ICF, for graphic illustrations of their developing systems.

Finally, it has to be acknowledged that the book provides only a personal glimpse at a vast topic. Acknowledgement has to be made to numerous exculsions that cannot be included due to limited space.

From this starting point the reader can collect and reference their own examples to futher the subject into a personal compendium of decorative and innovative applications of concrete.

Wᵉ are in an era when the visual potential of concrete as both an interior and exterior finish is being increasingly specified to support the wider drivers for the use of local materials, sustainability, material and energy efficiency and cost certainty.

This book provides a fascinating explanation of the material's history (from ancient Egypt to present day and beyond), and the technical background of both material and design solutions adopted by many of the key 20th and 21st century architects and engineers from all over the world.

As well as including a cornucopia of concrete exemplars, the technical information and insights will both inspire and inform industry practitioners and general readers alike and successfully lays to rest the 'grey ghosts' of accepted current public perception about concrete. The superficiality of this perception is highlighted by some of the well known and loved examples so well illustrated.

For those wishing to learn and to become well informed about the most used material on earth (apart from water) the author has provided an inspired, informative and technical amalgam which considering the vast subject has been transduced to allow a panoramic view of its potential.

Together with two other recent publications *Sustainable Concrete Architecture**and *Concrete: A Studio Design Guide**, *Decorative and Innovative Use of Concrete* will provide any student of architecture or engineering or design professional with the wherewithal, inspiration and justification for meeting the challenges of the 21st century with the use of this fascinating material.

Guy C W Thompson BA BArch RIBA
Head of Architecture, Housing and Sustainability
The Concrete Centre

*RIBA publication

FOREWORD

Thisbook explores how concrete can be seen as decorative and/or innovative, or cast or formed into artefacts that appear so. Persuading the reader that concrete can be decorative or innovative might be considered difficult when so much of what we see around us exhibits poor design or inappropriate use of the material – for example, poorly designed underpasses covered in tasteless urban graffiti; hard to negotiate car parks dirtied by pollution; elevated motorway routes running through city landscapes bringing with them noise and pollution – the so-called 'concrete jungle'.

These examples illustrate how the majority judge concrete from what they observe in their city locality. The conclusion formed, and perpetrated by the media, is that concrete intrinsically degrades our environment. However, the afflictions in the examples above have nothing to do with concrete as a material, but rather are a result of how it has been used.

Other examples come readily to mind: tower blocks having to be demolished due to poor insulation; damp interiors caused by interstitial condensation; elevated roadways in constant need of repair; grey concrete monoliths on rundown estates breeding crime and bringing despair to the inhabitants. Once again concrete itself cannot be held responsible; it is the way it has been used. There is no doubt that the 1960s left a poor legacy of questionable uses of concrete; however, whatever material had been used the outcome would have been the same due to the flaws in design and poor construction practices prevalent during that period.

So why might you be reading a book on decorative and innovative uses of concrete? Out of curiosity possibly, but

Introduction

CHAPTER 1

more than likely because there is within you a glimmer of hope that the material could be used to provide decoration rather than just structurally intrude into our lives. You also, presumably, have an interest in considering concrete as decorative or capable of being made to appear decorative.

Some architects and engineers set out to design concrete structures without aesthetic consideration since they intend to clad the concrete with other materials. Others strive to produce concrete with a particular finish, copying or mimicking some building or artefact produced by those whose work they admire, but find that their creations fail to perform a decorative role – often due to lack of attention to weathering characteristics. It is those who combine flare and understanding who succeed. This book sets out to trace the history and current output of such exponents.

All too often concrete suffers from the publicly held opinion of being 'grey and boring', exacerbated by the media proclamations of 'concrete jungle' which seem to be rolled out again and again. It is hoped that, by example, this book can change that belief into an appreciation of the successes of the material, whilst also shedding some light on the reasons for the failures.

The book gives examples of concrete as used in structures, as well as other assorted uses in which it has been made to appear decorative or which are intrinsically innovative. A selective history of some well-known iconic buildings is presented, together with some other less famous applications that arguably deserve more recognition. Various techniques that can be used to texture or colour, or just form or alter, the surface of concrete (plain or otherwise) into a more pleasing decorative surface are explored.

I hope this book will succeed in convincing the reader that concrete is no more or less intrinsically decorative than say stone, provided it is used in appropriate complementary designs. For example, it can be decorative in simple plain white surfaces provided the design, proportions and usage are complementary and include consideration of the weathering characteristics of the concrete. Some concrete benefits from post-cast treatment such as colouring with stains or abrasion of the surface to provide a texture; this is considered by many as the essence of decorative concrete, but is in reality only one small part. In cases where a disappointing finish has been produced or a change in use of a building requires a more suitable external surface, it is hoped this book will provide inspiration and guidance.

CHAPTER 2

A major difficulty in any book that deals with decorative and innovative uses of concrete is that the topic is subjective. No rigid rules can be applied. There is no universal definition of 'decorative concrete'; each person has their own way of judging whether a building, building element or concrete artefact is decorative or not. A designer may have wanted his building to appear 'distinctive' but for others viewing it this may amount to 'ugly'. Success in the use of concrete in a decorative format requires a combination of correct choice of materials, skill and flair in application and an appreciation, through design, of the weathering of profiles to ensure durability in service.

This book aims to encourage use of concrete in a more considered way, other than just as a structural mass. The examples illustrate how others have made use of the latent potential of concrete, and it is hoped they will stimulate new and repeat uses that will help to transform the public misconception of concrete being grey and boring.

That is for future chapters. What of the present, and how most think of decorative concrete? Depending on one's experience of concrete or how one chooses to view the material, decorative concrete can fall into one of two classes:

- Concrete that was decorative from its inception.
- Concrete that has been made decorative by some post-applied treatment.

A definition of 'innovative use of concrete' is the use of the material in applications that at first seem outlandish but on consideration are exploiting specific properties of the material. Thin sections can be made to provide chairs, helical stairways and boats using formulations that can be ten times stronger than 'normal' concrete. We even have concrete that is transparent.

This then poses the question: Can concrete be decorative as struck from the mould?

In one account, a plain formed concrete surface, the explanation of the finish by the designer is: 'High cost of preserving the landscape dictated that many architectural details in this complex were left intentionally rough'. I personally find it hard to consider this as a successful attempt at producing a plain finish decorative surface.

Examples of plain formed concrete finishes all in accordance with that definition in the Concrete Society Technical Report No. 52. Examples on pages 232, 237 and 277 have been cast against proprietary steel shutters, while page 248 [top] is from a profiled mould. The appeal of these examples is again in the eye of the beholder. It may be that the thin curved shell structure has the greatest appeal, but the type of finish shown on pages 237, 279 and 280 are currently in vogue for modernist concrete buildings. Which do you think are examples of decorative concrete and is the decorative property due to the location, shape or design, or the way light falls on the surface producing shadow and contrast; or is it due to the uncharacteristic use of concrete?

Some may believe concrete can only be decorative when it does not look like concrete. The example on page 192 has been vividly coloured. Examples on page 205 are thin artistic wall hangings and play on texture. All might be considered decorative in this context by not resembling 'normal' concrete.

Consider the decoration based upon a sketch by Picasso (page 9). We probably accept this due to the interest it gives us in determining just what it is and the material or structure is of no consequence. Any material could have been used: metal or plastic for example, or from this view it could even be a painting. Because we cannot determine this is concrete, we cannot apply any pre-conceived opinion on the make-up or structure and therefore no time is spent rejecting the image. Perhaps this illustrates a diametrically opposite impression to that created by the Morbio Inferiore on page 222.

The church walling on pages 130 and 131 is also well camouflaged and thereby captures our attention, in this context. One assumes stone as being universally acceptable as cladding so if concrete looks like stone it is automatically accepted. Stone is what comes to mind when one considers paving, plant troughs, flint walling, ashlar walling, and Cotswolds stone roofing tiles, but these can all be made of concrete.

The Colosseum (page 14) could be stone, but it isn't. As could the cobbles in Sheffield (page 160). Light passing through the wall on page 290 would suggest that it is not made of concrete – but it is!

Brilliant design will always grab the attention, whatever the material used. The striking columns designed by Frank Lloyd Wright (below) are so significant because they take the material into a new, erstwhile unimagined, structural arena, the slender concrete shafts rising up high and flaring

[Above] Concrete paving in a circular amphitheatre layout. The quality has a direct relationship to the weathering and durability characteristics of concrete. Carnfunnock amphitheatre (photo: Jack McGeown)

[Opposite] Upside-down columns at the Johnson Wax Headquarters, Wisconsin; Frank Lloyd Wright (photo: Bjørn Lund Mogensen)

What is Decorative Concrete?

[Above] The curved roof of *Canary Wharf Station, London (photo: David Groom)*

[Opposite] *The 'Y' building, Askersgata, Oslo. 'The Fishermen' is a wall decoration from 1970 made by Norwegian artist Carl Nesjar and based on drawings by Pablo Picasso (photo: Hans Nerstu)*

out to produce a roof, as well as allowing through plenty of light. The novel design won the day and concrete was the beneficiary. At that time it would have been expected to be steel or stone, or anything but concrete.

In a similar vein is a modern slender column and curved roof design (see opposite), now part of the award-winning railway station in London's Docklands. I was professionally involved with this building and remember it because the exposed as-struck concrete surface was initially considered unsatisfactory. Perhaps concrete has now become accepted for such use in structures. We no longer marvel at the performance but familiarity with the material's use in other applications has perhaps made us want more from the finish than can be provided by an as-struck form.

Plain as-struck concrete surfaces are not liked by some because the concrete can be seen to be concrete. Concrete is sometimes seen as a poor surface finish. It is only when it is treated to change the appearance that it becomes acceptable.

Acceptable concrete does not look like concrete. This is not my own belief but it tends to be the generally held opinion of the majority of those who judge or form opinions on the subject of concrete finishes.

Concrete as just a plain surface with the commonly held characteristic appearance of 'looking like concrete' can be 'improved' by grit-blasting a 'Picasso' on the surface. The surface is still the same surface; however, the eye is now drawn to the interesting sketch almost as if it were a kind of graffiti.

We all walk over acres of block and paved areas (page 7) and think of it as bricks or stone, anything other than concrete. Because we see so much of it around, in all those geometrical patterns and colourings, we tend to regard it as a version of the granite blocks common in the back streets of our old towns. All very acceptable.

Now a competitor to block paving has appeared: imprinted concrete. This has become popular because it is made to look other than like concrete – for example like riven stone or perhaps brick. Timber decking, as a bridge deck, is another clever camouflage to hide the concrete appearance and we find that surprising.

The art (page 205) and floor decking (page 170) would not be recognised as concrete. Images tend not to be formed in concrete (pages 276 and 277) and concrete is not usually blue or red and black – and what about coloured chairs (page 287)!

The answer is that concrete can be all or none of these. It can have vivid pigmentation but it does not need to be cast or painted so dramatically in order to be decorative or even innovative.

Decorative, by definition, means 'adoring; suited to embellish'. I would suggest concrete can either be made decorative or is decorative without post-applied treatment. The appreciation is in the eye of the beholder. I would argue that it is the skill of the designer, user or applicator that makes concrete decorative.

Concrete when bound together by hydraulic cement is intrinsically a stone-like material that can be formed into artefacts that resemble stone. It therefore should be considered as such – a man-made stone at one time given the descriptor 'flowstone'. Portland cement took its name from the fact that when hydrated and hardened it resembles Portland stone. Apart from the 'dark satanic mills' commonly built from stone, the use of stone for most statues, buildings and other general applications is readily accepted by the public irrespective of appearance; but concrete is too often condemned as grey and boring without further consideration. Most cannot tell the two surfaces apart when they are laid side by side.

There are forms of concrete bound together by binders other than Portland and hydraulic cement, and some of these have been used with great success to produce 'decorative artefacts'. This book includes examples of concrete with binders other than Portland cement since such material has been available for some time and the reader should be made aware of them.

Concrete materials with strengths far greater than could have been conceived only a few years ago have recently been developed. These appear to have extremely durable composition, resist various forms of ingress and allow thin sections to be loaded, and are suitable for structural applications. They are beginning to appear in decorative and innovative applications worldwide.

The appearance of decorative concrete can be influenced by five factors: texture, pattern, colour, form and weathering characteristics. The first three can be cast into the concrete or applied afterwards. The last two are a function of the design and usage.

This book will help the reader to discover those applications that they consider to be decorative and innovative. It is important to allow the inherent nature of the material to be used to its optimum rather than just for casting copies of other materials. Concrete is a compound, a compacted mass of coarse and fine aggregate cemented together by hydraulic or other mortar. It therefore has to be seen as such and used in that context with knowledge of its properties and requirements.

There have been – and continue to be – those who choose to use concrete without sufficient knowledge and thereby miss the opportunity to exploit the potential of this material in decorative and innovative ways. It is hoped this book will help to address that shortcoming by providing a better understanding of the capabilities of concrete.

Some of the oldest concrete so far discovered can be considered as decorative. Around 5600 BC a form of red lime concrete incorporating a mixture of lime, sand, gravel and water was laid to form floors for huts on the banks of the river Danube at Lepenski Vir in Yugoslavia. The material was evidently brought from 200 miles upstream.

3.1 Egyptian concrete

The earliest known form of Egyptian concrete, found at Thebes, dates back to 1950 BC. It is a mural depicting the method used by the Egyptians for manufacturing mortar and concrete. From Egypt, the technology spread around the Mediterranean and apparently reached Greece around 500 BC.

The Greeks also used lime as a binder for rendering walling. The palaces of Croesus and Attalus benefited from this application as well as a lime bedding mortar between bricks and stone.

3.2 Roman examples

It is thought the Romans copied the process of making concrete from the Greeks. The Latin word 'concretus' means 'grown together'. Examples of Roman concrete dating back to 300 BC have been found.

The 2nd century BC saw a significant leap forward. Until then Roman concrete had been in essence lime/sand mortar that hardened slowly by carbonation, converting the lime into calcium carbonate (limestone). At that time a source of pink sand was identified in the Pozzuoli area. This sand formed a much harder concrete, gaining strength quicker than any other concrete they had so far produced.

The sand was not an inert material as normally used, but a fine volcanic ash rich in silica and alumina that reacted with lime to form a silicate hydrate, now known as pozzolanic cement. Vitruvius exploited this material, and in his handbook for architects described it as being capable of hardening both in air and under water. We still use these materials today under the banner of hydraulic lime mortar; indeed, there has been a resurgence in their popularity due to their ability to more readily accommodate movement within masonry than Portland cement mortars. The first major application of pozzolanic cements by the Romans was in the core of walling of the theatre at Pompeii, 75 BC, where the walls were faced with stone and bricks.

The Romans failed to reinforce decorative concrete successfully, but did try by incorporating bronze strips into some constructions. The significant difference in thermal movement between the two resulted in cracking and spalling. Structures had to be designed to accommodate loads in compression and massive walls were needed to carry the loading, some up to 8 metres thick.

Methods of lightening structures were introduced that enabled the building of more delicate and decorative structures. These included casting earthenware jugs into walls and incorporating lightweight aggregates such as pumice, a volcanic porous rock. By these means some graceful and dramatic structures were built.

3.2.1 Colosseum, Rome, AD 80

The Colosseum was erected by the emperor Vespasian (AD 69–79), and paid for from the booty acquired following the rebellious war with the Jews in Palestine. The architect is unknown. The location chosen was on the site of a lake in the garden of Nero's palace, resulting in the requirement for considerable subsurface drainage.

The Romans built many concrete amphitheatres but the Colosseum was by far the largest, with a capacity of 50,000 spectators. Its purpose was as a venue for morbid entertainment – of watching criminals being killed by professional fighters, sometimes accompanied by animals. It was the fact the building had been used for this purpose, especially involving the martyrdom of Christians, that eventually secured its survival after a period of depletion

[Above] *Colosseum, Rome,* AD *80 (photo: Dr. Wendy Longo)*

[Opposite] *Pantheon, Rome,* AD *127 (photo: Udo Wiesner)*

Decorative and Innovative use of Concrete

during which the stone was robbed by members of the church and aristocracy.

Concrete foundations under the outer walls formed a ring up to 13 metres deep. The inner foundations under the arena were a mere 4 metres deep but were arranged to coincide with each of the concentric walls. All this was placed over drains 8 metres beneath the structure.

The design is different from other Roman and earlier Greek buildings. It had been customary to combine rectangular rows of columns and beams and the inclusion of a triangular pediment. Here arches and vaults were used, through employing brick-faced concrete that enabled greater spans to be achieved. The huge downward thrust of the outer walling managed to accommodate the sideways pressure from the vaulting in the circular promenades. The overall stability was achieved by the circular plan of the building.

The chosen ratio or proportion of the period was 5:3. The unknown architect apparently chose the arena to measure 300 × 180 Roman feet (a Roman foot is equivalent to 29.6 cm). The width of the auditorium equalled the width of the arena and the height of the external façade.

It is believed that the total length was planned to be 660 Roman feet, and the width 540 Roman feet. This would make the perimeter 1885 Roman feet – enough to allow the inclusion of the 80 grand entrances. Entrance arches in Roman amphitheatres were 20 Roman feet wide with columns between of 3 Roman feet wide. The Colosseum actually received a perimeter of 1835 Roman feet ($80 \times 23 = 1840$), and the arena was adjusted to 280×168 Roman feet. This still achieved the required 5:3.

At the first level, the floors were made of marble or travertine (the same rock as used in the outer walls), whilst the walls were of polished marble slabs and the ceilings painted stucco. The present pockmarked appearance (page 14) is due to the robbing of the walls in medieval times for iron clamps.

3.2.2 The Pantheon, Rome, AD 27

The Pantheon in Rome was begun in 27 BC by the statesman Marcus Vipsanius Agrippa, probably as an ordinary classical temple of rectangular plan with a gabled roof supported by a colonnade on all sides. It was completely rebuilt by the emperor Hadrian sometime between AD 118 and 128, with further alterations in the 3rd century.

It is a circular building of concrete faced with brick, with a great concrete dome, 43 metres in diameter rising to a height of 22 metres above the base. The interior is lit only by the light flooding through the 8 metre oculus (or eye) at the centre of the dome. There is a front porch of Corinthian columns supporting a gabled roof with a triangular pediment. Beneath the porch are huge bronze doors 7 metres high, the earliest known large examples of this type.

The Pantheon was dedicated in AD 609 as the Church of Santa Maria Rotonda, which it remains today. It is the first temple to combine concrete construction – a technique in which the Romans were especially innovative – with the more conservative if decorative use of Greek classical styling.

Two factors have contributed to the success of the building: the excellent quality of the mortar used in the concrete and the careful selection and grading of the aggregate material, which ranges from heavy basalt in the foundations of the building and the lower walls of the dome through brick and tufa (a volcanic ash) to the lightest, pumice, towards the centre of the vault.

A further innovation is evident from the outside where the drum of the walls of the dome is seen to be strengthened by huge brick arches and piers set above one another inside the walls, which are approximately 6 metres thick. Further strength is provided by the upper third of the drum of the walls coinciding with the lower part of the dome. There are relieving arches over the recesses inside but all these arches were, of course, originally hidden by marble facing.

The composition of the Roman concrete used in the dome remains a mystery. An unreinforced dome in these proportions made of modern concrete would hardly stand the loading from self-weight due to the low tensile strength of concrete. However, the Pantheon has stood for centuries.

The Pont du Gard, France AD 150 (photo: © Csld/Dreamstime.com)

It is known that Roman concrete was made up of a pasty hydraulic lime – pozzolanic ash and lightweight pumice from a nearby volcano and fist-sized pieces of rock. There is some similarity to modern concrete but it is probably the way the concrete was compacted in very small amounts and tamped down to remove as much air and voidage as possible that has enhanced its tensile strength significantly.

The dome is the largest surviving from antiquity and was the largest dome in western Europe until Brunelleschi's dome of the Duomo of Florence was completed in 1436. The Pantheon is still a church in which masses are still celebrated.

The Pont du Gard, AD 150, with its relatively delicate structure and concrete aquaduct running across the top, may be considered in both a decorative and a structural context. If nothing else it is innovative.

3.3 Post-Roman period

After the Roman period there was little use of hydraulic mortar up to the 4th century. It was once thought that the Normans brought concrete back into Britain; however, excavations in Northampton have revealed shallow bowls in the ground constructed by Saxons which were used as concrete mixers. A central post acted as a pivot for a horizontal beam with suspended blades that evidently agitated a mix based on limestone and burnt lime.

Norman concrete was much like Roman, being used as infill in walls which was then clad in some decorative facing. In Reading Abbey the original stone facing to the walls has been lost, thus revealing the concrete core. Concrete was used in the Tower of London, as well as in castles, churches and cathedrals. Lime ash concrete screeds have been discovered, used to protect wooden floors in large halls such as the upper-storey floor of Little Moreton Hall in Cheshire, c.1580, where the original screed has been restored using the original type of material.

Plaster was used widely in Europe in the Middle Ages and gypsum employed for both internal and external decoration on timber-framed buildings. Hair reinforcement as well as a range of admixtures (including malt, urine, beer, milk and eggs) were employed to modify the properties. Hair presumably improved the bonding of plaster; some

Decorative and Innovative use of Concrete

of the other additions appear questionable but perhaps improved the workability of the plaster.

In the 14th century a form of decorative external plastering was introduced: pargeting. This became popular in the south and east of England and can be seen today on timber-framed buildings. The appearance is of a thick coat of lime putty or sometimes a mixture of lime and gypsum plastered to a deep-textured relief.

In Europe at this time terracotta was reintroduced and became the vogue for external ornamentation. Then in the 15th century a new type of render was developed by the Venetians. It was given the name 'marmorino' and consisted of lime applied directly onto masonry. A further development in the 16th century led to a highly decorative internal plaster called 'scagliola', invented by the Bavarian stuccoists. It was composed of gypsum plaster intermixed with animal glue and pigment and used to imitate marble. Sometimes marble dust, sand and lime were used to supplement the mixture.

Around this time the sgraffito technique, in which layers of contrasting coloured lime plaster were applied to building façades, was also developed. The technique was not new but had been practised by Italian artists working in Germany. Once the layers had been applied, the outer layers were cut or scratched through to reveal the lower layers in an intricate patterned form.

Various internal plaster forms resembling marble were developed, based on a scored grounding of lime or gypsum plaster with a thin overcoating of fresh lime or gypsum into which pigments were scattered. These were given the name 'stucco lustro' or 'stucco lucido'.

In the 18th century innovative plasters were developed that incorporated oils derived from, for example, tar, turpentine and linseed. In 1765 David Wark patented a 'stone paste', a lime-based mix including 'oyls of tar, turpentine and linseed'. Others were patented by the Rev. John Liardet in 1773 (his included a drying oil) and in 1777 by John Johnson. Many of these oil-based renders failed, with the result that water-based renders regained popularity. In 1779–80 Dr Bryan Higgins patented and wrote about a 'water cement, or stucco' consisting of lime, sand, bone ash and lime water.

3.4 Decorative concrete developments in the 18th and 19th centuries

Whilst the initially promising lime and oil-based renders and stucco were beginning to fail, the conditions were not yet fully established for hydraulic cement or hydraulic lime-based binders to take total control of the market. The interim period saw the development of 'artificial stone' products. These had the benefit of high early and long-term strength as well as a durable surface and internal structure. Two noteworthy examples are considered here.

3.4.1 Coade stone

Coade stone was manufactured between 1769 and 1843 by the two Coade ladies, mother and daughter, who both traded under the name of Mrs Coade and had the same Christian name, Eleanor. The origin and patent for the product rests with a Richard Holt, who started production in the early 18th century.

Coade stone is akin to terracotta. It is a fired form of kaolinite clay containing titanium oxide, feldspar, quartz and grog of previously fired material ground down for recycling. The feldspar acts as a flux and the quartz provides the glass-like consistency. The end product looks like stone but bears little physical relation to stone.

Production required a clay master and a mould made from it. The mould was then assembled and raw Coade stone grog poured into it. On drying pieces would be stripped from the mould and some larger pieces would be assembled from sections stuck together with a slip of liquefied clay. The material was evidently stable during firing and large thin slabs could be produced.

The skill and ability of this company is worthy of note. Workers had to fire these items at temperatures of up to 1100–1150°C and maintain them at those temperatures over four days.

Some notable sculptors were engaged in making Coade masters. Large pieces were sometimes assembled in the drying state using iron or steel brackets and dowels. Coade stone has proved to be durable; however, its one weakness has been corrosion, with consequential expansion of the fixings causing deterioration of some pieces.

Decorative and Innovative use of Concrete

Angel and Father Time figures on the Lady Henneker Monument, Rochester Cathedral, from the John Bacon era (photo: Geoff Matthews)

Eleanor senior and her husband George Coade began business in the woollen manufacturing trade in Lyme Regis but the general decline in that trade, accompanied by their bankruptcy, caused a move to London to St Thomas Apostles Street. Eleanor junior had already moved in with Daniel Pincot, a producer of artificial stone at King's Arms Stairs, Narrow Wall, Lambeth. It is not known how Coade stone became part of the business arrangement between Daniel Pincot and Eleanor junior, but it was not long before she took a controlling interest in the business. She sacked Daniel in September 1771, announcing the situation in *The Daily Advertiser*, *Gazetteer* and *The New Daily Advertiser*. This was followed two weeks later by an announcement that John Bacon would be superintending the business.

The Coade's name for their product was 'Lithodipyra', a combination of Greek words that can be broadly interpreted as stone-twice-fired. For about 60 years the factory produced a range of architectural ornamentation including figures, column capitals and heraldic door heads. The secret of their manufacturing process was guarded but nevertheless others set up in competition, with reputedly less satisfactory wares.

John Bacon died in 1799, after which John Sealy was made a partner in the company and the firm's stamp became Coade and Sealy. John Sealy died in October 1813 aged 64, and Eleanor senior engaged a William Croggon from Cornwall who managed the factory up until her death on 18 November 1821 at the age of 88.

William Croggon had expected the business to be left to him by Eleanor senior but in the end had to buy it. He went bankrupt in 1833 as a result of unpaid debts to the Duke of York, and he died in 1835. Croggon's son Thomas John re-founded the business and it traded until 1977, keeping the Coade factory premises but trading in manufactured goods. The Coade stone business itself continued in Lambeth only until 1843 when the moulds were sold to H.N. Blanchard who had served his apprenticeship with the firm. He already had premises in Blackfriars and work continued here until 1870. Between 1851 and 1875 a terracotta factory was operated by Blanchard, who was by then making Coade with a creamier appearance.

Many different commissions were undertaken, including a screen in St George's Chapel, Windsor. Outputs included many artefacts that might otherwise be carved in natural stone. During the time of John Bacon much Coade stone was supplied to Buckingham Palace. Eleanor Coade's reputation grew. Orders were received from throughout the country and from as far afield as Rio de Janeiro and Pushkin, St Petersburg. A statue of Nelson was supplied to Montreal after his death.

At the Royal Naval Hospital at Greenwich there is a memorial on the pediment left by Wren showing the body of Nelson being given to Britannia while England, Ireland and Scotland mourn. Completed around 1813, it is 12 metres long and 3 metres high and the condition is still excellent, confirming the durability of the product. The assembly was modelled by Panzetta and West and took three years to complete, at a cost of £2584.

Alison Kelly, author of *Mrs Coade's Stone* (used to compile these notes and illustrations), has had analysed a fragment of Coade stone from the decoration of Belmont, the house in Lyme Regis owned by the Coades. The analysis, carried out at the British Museum Research Laboratory, indicated the raw material to be ball clay from Dorset or Devon, to which was added 5–10% flint, 5–10% quartz sand, at least 10% grog (probably crushed stoneware) and about 10% soda-lime-silica glass, which acted as a vitrifying agent.

A notable example of Coade stone stands on the southern approach to Westminster Bridge; another larger one can be seen at the All-England Rugby Football Club at Twickenham. These were originally used to adorn the Lion Brewery on the South Bank, London – the larger one on the roof and the smaller one over the entrance.

In the 1950s only three Coade stone sites were known: Captain Bligh's tomb at St Mary's Lambeth, the Coade stone lion at Westminster Bridge, and the houses, all with Coade stone decoration, which survive complete in Bedford Square near the British Museum. Because it so resembles natural stone it is often difficult to identify. By the time of writing her book in 1990, however, Alison Kelly had identified 66 Coade stone artefacts.

Coade stone lion, Westminster Bridge, London (photo: Alistair Hall)

3.4.2 Ransome's stone

In 1856 the inventor Frederick Ransome patented an artificial stone he called 'siliceous concrete stone', which later became known as Ransome's stone. At the Patent Concrete Stone Company in East Greenwich he set about production of numerous items. The business appears to have flourished and the size of the works grew. The company produced large amounts of architecturally decorated blocks used 'in the erection of building both here and abroad'.

The product and process was so successful that in 1869 Ransome's son Ernest Leslie Ransome (1844–1917; born in Ipswich) moved to San Francisco and the following year set up the Patent Stone Company, initially producing the same artificial block products. Ernest Ransome went on to patent a form of twisted reinforcement that became the cornerstone for the 'Ransome system of reinforcing concrete' employed in many concrete structures in America, including the United Shoe Machinery Corporation building in Massachusetts, the Greystone Winery, California, and the Borax Warehouse, New Jersey. He is credited as having done more for reinforced concrete in America than any other engineer.

Ransome's stone was more akin to conventional concrete than Coade stone since the binder is calcium silicate. The manufacturing principle is described in two references of the period: *The New Guide to Masonry, Bricklaying, and Plastering: Theoretical and Practical* (edited by R. Scott Burn, 1871) and *The Manufacturer and Builder*, Vol. 2, Issue 6 (June 1870).

In summary, sodium silicate (known as water glass) is mixed with sand (and sometimes gravel, flint, chalk or lime-stone) in an aqueous solution of sodium hydroxide (caustic soda). The caustic soda dissolves the flints and the solution takes on a treacle-like consistence as it cools. The solution is then compacted into the moulds. It is then carefully de-moulded and washed over with a cold solution of calcium chloride. The surface rapidly hardens due to the reaction between calcium chloride and sodium silicate which forms insoluble calcium silicate and sodium chloride (common salt). The piece is then placed in a boiling solution of calcium chloride until saturated and the conversion completed. The salt is then washed away and the piece left to dry.

Decorative and Innovative use of Concrete

The following account from the 1870 issue of *The Manufacturer and Builder* mentioned above gives details of the process, 'as now executed in East Greenwich':

The first operation is the drying of the sand. It is for this purpose lifted by means of an elevator (similar to our grain elevators) and discharged into the upper end of the inclined rotating cylinder of galvanized iron, through which a current of dry hot air is passed upward by means of a fan-blower. The sand is then passed through a sieve and preserved for use in closed chambers, as the natural sand would often, by the large size of its particles, produce a stone of coarse grain size; a portion of it is crushed between cast-iron rollers, which are also used to pulverize the limestone.

In one section of the works are lime furnaces in which the silicate of soda is prepared. They are similar to glass furnaces, and contain pots in which flint or quartz is melted together with carbonate of soda, so as to produce a glass which will more easily dissolve in the boiling solution of caustic soda than the unprepared quartz. If fifteen parts of flint are melted together with eight parts of carbonate of soda, a compound is obtained which will melt in boiling water, and which is the so called water glass; but if less soda is used, the temperature of boiling water is insufficient to accomplish the solution, as greater heat is then required. This is done in the Ransome process; the amount of soda is made as small as possible; therefore the boiling of the thus prepared soda glass with the caustic soda solution is accomplished in cylindrical boilers, similar to steam boilers, in which the pressure is brought to about seventy pounds per square inch, corresponding with a temperature of 396 deg. F. This is sufficient to dissolve the maximum amount of flint in the caustic solution, and to form a liquid rich in silex. In fact, the value of this preparation depends on the smallest possible amount of soda present.

There are two ways of securing this; the first may either be first melted with a sufficient amount of soda, 15 to 20 per cent, so as to make the resulting glass soluble in water of 300 deg. or more, or the unprepared flint may be placed in a strong caustic soda solution, and submitted to a prolonged heat of this temperature. The latter way takes more time. When drawn off from time boilers, the liquid has a specific gravity of 1.2, and after clarifying by settling, it is concentrated by evaporation to the consistency of syrup and a specific gravity of 1.7. It must then be perfectly transparent and gelatinous.

The mixing of this substance with sand is accomplished

by means of a mill with cast-iron disks. The material ground is: coarse sand, finely powdered sand, powdered limestone, and the silicate of soda powdered as above. They are ground up to a perfectly homogeneous mass in the short time of 3 minutes for every charge, which consists usually of 18 litres of dry material to 1 of the liquid; but sometimes for special purposes as much as 24 litres of the dry material are used to one of the liquid. The mixture is perfectly plastic, and possesses just enough cohesion to be put into moulds. These must be well greased to prevent adhesion, and strong pressure is absolutely required in order to prevent porosity of the product. Thus far, the operation of placing the mixture in the moulds is done by manual labour, but will soon be performed by proper machinery.

The next process is the hardening of the objects. The removal from the mould must be accomplished with some care, as there is still little cohesion; but under the influence of the solution of chloride of calcium, they obtain in a few minutes sufficient hardness to be handled and transported without peculiar precaution. Formerly the large objects were plunged into a bath of chloride of calcium, in order to impregnate them; but lately the same effect has been reached in another very interesting manner. During the moulding in the form, a hole is made somewhere reaching to the centre of the object, and in this hole a tube connected with an air pump is inserted, which is closely fitted into the entrance of the hole; if then, after removing the object from the mould and plunging it into the solution of chloride of calcium, a vacuum is caused by the air space, the liquid will be very quickly absorbed and it thus penetrates the artificial stone to its very centre. To treat flat stones in this way, iron moulds are made with perforated bottoms, which are placed in proper position, and the air under their bottoms is exhausted; atmospheric pressure will then drive the liquid entirely through the upper side. This method of exhausting the air from the inside or from below large objects, in order to enforce a perfect penetration of the desired place, gives very rapid and highly satisfactory results. Small objects are simply moistened by pouring the liquid over them.

After the hardening with a cold solution of chloride of calcium, follows a bath of the same liquid of a specific gravity of 1.4, which is heated to the boiling point by means of steam tubing. In this way, all air is perfectly removed from the pores, and the chemical affinity between silicate and chloride increased. In East Greenwich is a row of such hot baths, at one end of a small railroad, on which the objects are continually and slowly arriving; and after passing through these hot baths, they enter a series of douche baths, intended to remove the chloride of sodium previously formed. These double baths are chambers with perforated bottoms and tops and the water continually passing through them is used over and over again, until it contains too much salt, when fresh water is substituted. The operation is continued till all the salt is perfectly removed; then the objects are dried, as much as possible by the usual summer heat, and during the winter in drying chambers which have a temperature of 70 or 80 deg. F.

The artificial stones of Ransome distinguish themselves by the sharpness of their forms, equal colour, resistance to heat, frost, dirty water, all atmospheric influences, and, finally and chiefly, by their cheapness. A large mass of objects, from the most complicated, compete now with those made of natural stones in England, the Indies, and on this continent.

Production continued both in England and in the USA under the supervision of Frederick and Ernest Ransome. A curious use of the stone was as drinking water filters. There were even decorative versions made of these, for sitting on sideboards in the drawing room. A filter 7 inches in diameter by 1 foot 4 inches tall could process 5–6 gallons per day, and cost 10s. 6d. They were also used in commercial premises, and many learned authorities and individuals gave testimonials praising the products' ability to filter out all sorts of particles and contaminants. It was claimed that the porosity and filtering properties could be tailored in the manufacturing process.

At this time concern was growing regarding the rapid decay evident on the external stone elevations of the new Houses of Parliament. The causes of decay in the stone were attributed partly to mechanical reasons, but according to a Professor Ansted 'may generally be traced to the absorption of water'. Another eminent professor, Sir Edward Frankland (1825–99), a pioneering chemist and 'inventor' of the chemical bond (he was known as the father of valency), served on the committee appointed to examine the cause of the decay. He also instituted a series of experiments in connection with Ransome's stone. The chief object of these was to expose the samples to similar influences to those experienced by stones used in outside contexts in larger cities at that time. Samples of Ransome's stone were tested alongside the natural stone that was in common use at that time.

The programme used the following accelerated test procedure. First the samples were tested to determine the porosity by water absorption. Each sample was then subjected to 48 hours immersion in three concentrations of sulphuric acid. The samples were then boiled in water to remove all the acid and brushed to remove any loose material. The results are shown in the table opposite.

The Ransome's stone used in these tests was just 14 days old and it was stated that it would become harder and more crystalline by natural aging. However, from the results presented Professor Frankland concluded:

> Mr. Ransome has invented a material which, with the exception of granite and primary rocks, is better capable of giving permanency to external architectural decorations than any stone that has been hitherto used.

3.4.3 Ransome's stone and the Bolinder Palace

I visited this palace in 2010, that is now part of the Grand Hotel, Blaisieholmen, Stockholm.

Jean Bolinder (1813–1899) was a mechanical engineer and interested in the latest methods of construction. He built this – his residence – between 1874–1877. He had widespread interests throughout Europe and with his architect, Helgo Zettervall, chose Ransome's Artificial Stone to decorate the façade.

Details included in the masonry for the brick walls can be seen on pages 30 and 31. A restoration was undertaken in 2002 when the stone, then 125 years old, was investigated and found to have a density of circa 2250kg/m³, compressive strength of 73Mpa and the carbonation front had only penetrated 5 to 7mm.

The investigation of the Ransome's Stone was reported by Bo Nitz of Optiroc AB, P.O.Box 707, SE-169 27 Solna, Sweden, who undertook the investigation and provided an account in 'Microstructural analysis of the render of the Bolinder Palace in Stockholm' Materials Characterization. Vol 53, Issues 2–4, November 2004, pages 187–190. EMABM 2003: 9th Euroseminar on Microscopy Applied to Building Materials.

Decorative and Innovative use of Concrete

Name of stone	Porosity (% of water absorbed)	Per cent alteration in weight by immersion in dilute acid						Total % loss by action of acid and subsequent boiling in water	Further loss by brushing	Total degradation from all causes
		Of 1%		Of 2%		Of 4%				
		Loss	Gain	Loss	Gain	Loss	Gain			
Bath	11.57	1.23	–	2.82	–	2.05	–	5.91	0.26	6.17
Caen	9.86	2.13	–	4.80	–	0.67	–	11.73	1.60	13.33
Aubigny	4.15	1.18	–	4.00	–	–	1.04	3.56	0.29	3.85
Portland	8.86	1.60	–	1.10	–	1.35	–	3.94	0.24	4.18
Austen	6.09	3.52	–	3.39	–	3.11	–	11.11	0.27	11.38
Whitby	8.41	1.07	–	–	0.53	None	None	1.25	0.18	1.43
Hare Hill	4.31	0.75	–	–	0.60	None	None	0.98	0.15	1.13
Park Spring	4.15	0.71	–	–	0.10	0.15	–	0.81	None	0.81
Ransome's Patent	6.53	–	9.5	None	None	None	None	0.63	0.31	0.94

Comparative results of testing Ransome's stone and natural stone

Some of the façade stone was applied as a render on site. Such an application of Ransome's stone is not documented elsewhere and warrants further research along with finding other remaining examples of his output. Publications of the late 1800s suggest Ransome's Stone was widely in use, but few examples are now known. The Bolinder Palace must certainly be one, if not, the most impressive example of the product and an illustration of its longevity, especially in both a marine environment and one repeatably subjected to freeze/thaw in winter.

3.4.4 The run-up to Portland cement

The failure of oil-based renders combined with the need for more durable binders brought about a revival in hydraulic limes. This coincided with a need for mortars and concretes that would not be adversely affected by water. John Smeaton, an English engineer, had a particular use for a more durable concrete. He was commissioned to build a replacement lighthouse on Eddystone Rock, about 15 miles southwest of Plymouth.

Previous attempts had used timber. One had burnt down and the other had been demolished by a storm. Smeaton realised it would be prudent to build using stone but that this would require a binder that would set relatively quickly and withstand the action of seawater (until then the cements available were weak and slow to set). Smeaton conducted some investigations and in 1774 found that quicklime made harder cement if combined with clay. Lime mortar did not provide the complete answer so he investigated further

[Below and across] *The Bolinder Palace, now part of the Grand Hotel, Blaisieholmen, Stockholm (photo: Author's collection)*

31

A Brief History of Decorative Concrete

and arrived at a mixture of burnt Aberthaw blue lias, a limestone from South Wales, and an Italian pozzolana from Civitavecchia. When the two were combined they produced cement that had superior properties when placed to harden under water. The resulting concrete was the most durable since that of the Romans in the 5th century and represented a huge leap forward in concrete technology. Smeaton wrote about these developments in his book *A Narrative of the Eddystone Lighthouse*, but did not present his findings until 1791, long after he had finished the lighthouse.

The lighthouse was completed in October 1759. In 1876 it was discovered the sea was undermining the rock beneath it, so it was taken down and replaced by a larger structure nearby. Smeaton's lighthouse was rebuilt on Plymouth Hoe, where it stands today, and the concrete stump remains on the rock where it was cast around 250 years ago.

After Smeaton's book became available various cements were patented that apparently applied some of his findings. In 1796 Rev. James Parker patented 'Parker's Roman cement', a hydraulic cement produced by calcining naturally occurring nodules of impure limestone containing clay. When mixed with sand it could be used for stucco and could also be cast to form mouldings and other ornaments. These nodules, septaria from the London clay, were found on the beaches of the Isle of Sheppey and the Thames estuary. The cement produced an unattractive brown-coloured mortar and needed a surface-applied finish. Parker sold his patent to James Wyatt and his cousin Charles Wyatt and then went to the USA. The patent lapsed and from about 1810 others began to manufacture Roman cement. One of these was the Medina works on the Isle of Wight.

Although various types of lime mortar concrete were in use from about 1820, the Medina Cement Co. on the Isle of Wight claimed to be the first (from about 1840) to produce a concrete in commercial quantities, based on Roman cement. Roman cement requires skilful mixing and sets quickly; it can only be made in small batches with the result that construction is slow. The first houses to be built of concrete (as opposed to concrete blocks) were at East Cowes in 1852. William Jessop used this type of cement on the West India Docks in London, one of the first structures to use concrete on such a large scale.

[Left] *John Smeaton who built the Eddystone Lighthouse using his form of quick setting cement (produced using limestone containing clay).*

[Below] *Statue of the Prophet Samuel in infancy, c.1850. One of the earliest concrete artefacts attributed to James Aspdin (photo: Edwin Trout, Information Services, The Concrete Society)*

A Brief History of Decorative Concrete

In 1811 James Frost took out a patent for an artificial cement that incorporated ground calcined chalk and clay. Between 1812 and 1813 the French engineer Vicat experimented with calcined synthetic mixtures of clay and limestone and in 1818 finally introduced a product. In 1822, back in the UK, James Frost patented another cement, similar to Vicat's, that he called 'British cement'; then came along Joseph Aspdin.

3.5 The invention and development of Portland cement

Joseph Aspdin (1788–20 March 1855) was a British mason, bricklayer and inventor. He patented Portland cement on 21 October 1824. The eldest son of a Leeds bricklayer, he began using artificial cements made by burning together ground limestone and clay. He named it 'Portland cement' because he thought it resembled Portland stone. Aspdin's cement was undoubtedly the most superior of its time; however, soon after its introduction developments in the

[Opposite] *The Wakefield Arms, Kirkgate, Wakefield, 1843–5. The only known building to have survived, brick built and rendered with Joseph Aspdin's Portland cement (photo: Edwin Trout, Information Services, The Concrete Society)*

[Below] *At Swanscombe, Kent, the first all concrete house (photo: Edwin Trout, Information Services, The Concrete Society)*

William Aspdin's Patent Portland Cement Works, Gateshead, c.1852. This was the largest cement works in the world at that time (photo: Edwin Trout, Information Services, The Concrete Society)

manufacturing process started to take place, such that present-day Portland cement has only the name and raw components in common.

Aspdin set up his first cement works at Kirkgate in Wakefield (1825–38) in partnership with William Beverley (1826–28), and then had to build a new works in Inge Road, Wakefield in 1843 after the railway extension took part of the site. The following year he retired and the business was taken over by his first son, James. So far only one surviving building has been identified that incorporates cement from that first works, namely the *Wakefield Arms*, a brick-built structure near Kirkgate Station, faced on the outside with a rendering of Portland cement. It is very close to the then site of the cement works.

Whilst the new works in Ings Road, Wakefield was in operation, Joseph's youngest son, William, left home for London. In 1841 he set up his own cement business in Rotherhithe, producing cement that was chosen by Sir Marc Brunel for his Thames Tunnel. This is thought to be the first major use of Portland cement. In 1847 William moved to Northfleet and set up in partnership with two businessmen at Northfleet and Swanscombe (these sites later became part of the Blue Circle corporation). At that time cement was sold in barrels; one delivery that had sailed aground at Sheerness on the Isle of Sheppey was used to build a public house, *The Ship on Shore*.

At this time attempts were taking place to employ cement in a decorative form. William Aspdin tried promoting the use of concrete for house construction, in 1850 purchasing land overlooking Gravesend in Kent and setting forth to build a mansion known as Portland Hall. It was abandoned half completed after nearly £40,000 had been spent on it. A small section of walling remains with precast concrete capping units, probably some of the earliest commercially cast.

In 1852 William moved to Gateshead-on-Tyne where he set up what was probably the largest cement works in the world at that time, with a production capacity of 3000 barrels (c.650 tonnes) per week. In 1860 he also started cement production in Germany, where he died four years later. The biggest drawback in using cement at that time was the relative cost of manufacture (about 10 times that

of today). With the development in the 1880s of rotary
kilns (for example as designed by Ransome and installed at
Arlesey near Hitchin, Hertfordshire) the cost started to come
down, but not before further developments in the USA had
been introduced back into the UK.

Isaac Johnson improved the production process by
raising the burning temperature so as to clinker the raw
feed. This produced much stronger cement, more akin
to that used today. In 1856 Johnson, who had a works
at Northfleet, took over Aspdin's abandoned works at
Gateshead. He also built a house at Gravesend, probably to
show off the potential of Portland cement.

One significant advantage of Portland cement over
pozzolanic and lime-based products is its relatively
quick (and consistent) set and gain in strength, enabling
construction to proceed at a faster rate.

3.6 Decorative use of concrete from 1840 to 1940

A significant early application in this period was the use of
Portland cement render on the new Osborne House on the
Isle of Wight. The builder was Thomas Cubitt, but evidently
Albert the Prince Consort designed the house. Concrete
was also employed in constructing the house. Such uses
gave Portland cement credibility; however, production
at that time (1845–48) employed vertical bottle kilns, so
manufacturing costs were still relatively high.

In 1873 decorative concrete panels were set into the
wall elevations of a large house, Down Hall, at Hatfield
Heath in Essex. This provides a rare example of the use
of concrete at that time on such a prestigious house. Of
note is the way the panels have weathered, adding to the
appearance of the house. At about the same time William
Lascelles patented a timber-framed domestic housing system
to which R. Norman Shaw applied external and internal
designs (see below).

3.6.1 Further use of Portland cement concrete in housing

In the early part of the 19th century it became apparent
that Portland cement concrete required reinforcement in
the zones intended to accommodate tensile forces. Various

A precast concrete house, Croydon, 1882. Concrete panels are fixed to a timber frame. William Lascelles patented this system. (photo: Edwin Trout, Information Services, The Concrete Society)

attempts were made with the inclusion of lattice iron tie rods (James Frost took out a patent in 1822 for including iron in concrete), while the French engineer Joseph Monier cast plant pots incorporating reinforcement and François Coignet encased iron frameworks in concrete. However, the person who is generally considered the inventor of reinforced concrete is an unknown builder from Newcastle, William Wilkinson, who registered a patent application in 1854 in which he proposed the introduction of strips of hoop iron at discrete distances apart depending on the strength requirement. He also proposed the use of secondhand wire colliery rope with the ends splayed out to provide a mechanical link to prevent slippage under flexural loading. The evidence of Wilkinson's patent came to light when one of his buildings was demolished in 1954. He had used a form of plaster-permanent formwork with 10 mm wires laid between the units and plaster and concrete poured over to encase the floor then topped with a granolithic slab. The wire was found to be in good condition after nearly 90 years in service.

In 1866 a Joseph Tall built cottages in Bexleyheath, Kent that included a lattice work of hoop iron embedded in the concrete floors. He had taken out a patent the previous year. Although today the cottages might not be considered as beautiful examples of decorative concrete, they did bring the use of Portland cement down to the working class level of building, incorporating surface finishes similar to the appearance of the more pretentious structures such as Osborne House and Down Hall.

Truly decorative concrete finishes were achieved in the system of timber-framed housing patented by William Lascelles in 1875. Examples survive in Sydenham Road, Croydon, where precast concrete panels measuring 1 metre × 0.75 metres and 40 mm thick, pigmented by iron oxide, have now provided a weathered terracotta appearance; the finish was originally disliked when new as it was considered too bright. The internal ceilings have precast slabs of a floral pattern typical of the Arts and Craft movement of the time.

Another of Lascelles' houses, located in the grounds of the Royal Horticultural Society at Wisley, Surrey, 1890, has external precast concrete panelling that resembles terracotta tile cladding.

New Jerusalem Church, Crystal Palace, London, 1883 (photo: Edwin Trout, Information Services, The Concrete Society)

3.6.2 Portland cement concrete introduced into church buildings

As acceptance of the use of Portland cement grew, churches also began to be built in concrete. One of the earliest examples was a Presbyterian church in the village of Loanhead near Edinburgh, constructed in 1875 in cast stone. Glanville's *Modern Concrete Construction*, dated 1939, reports the condition of the work as being

> good, the surface clean and free from noticeable crazing. The effect of weather has been so slight that the chiselled pattern impressed in moulding is still practically perfect. There are some large cracks in the walls, but these are due to coal-mining subsidence. The stone was made by local labour and cannot be said to be of the highest quality according to modern standards [1939] for the aggregate contained a proportion of very inferior material – shale from burnt-out coal-shale bing – and some surface defects have occurred from this cause. A rather interesting feature is that, under the arch of the doorway the cast stone is blistering in a manner similar to that sometimes seen on natural stone in a like position, showing that the weathering properties of the natural and synthetic materials have something in common. During the long period of weathering the stone has developed lichen growths and generally, the building presents a very natural and pleasing appearance.

In 1883 the first concrete church to be constructed in London, the New Jerusalem Church, was erected near Crystal Palace. It was built in mass concrete with walls 0.75 metres thick and pigmented red, at a cost for the design and erection of £3000. The church magazine of the time, the *Morning Light*, mentions the Patent Concrete Building Apparatus Company which played a part in the construction.

3.6.3 Edison and his involvement with Portland cement and concrete

Many will know the prolific American inventor Edison for his invention of the incandescent bulb, or light bulb as we now call it. He introduced a total of 1093 US patents, of which 49 involved or were related to cement and concrete. In 1899 Edison founded the Edison Portland Cement

Company, using leftover heavy equipment from a futile iron-ore refining enterprise.

Edison's involvement at a time when concrete was rarely used helped to bring the product to the fore. His cement manufacturing plant went onstream in 1902. Portland cement had become much cheaper since the development of rotary kilns by Messrs Hurry and Seaman in the USA. It is worth noting that the first rotary kiln was patented by Frederick Ransome in 1885, with trials being carried out at Barnstone in Nottinghamshire, Grays in Essex and Penarth in South Wales. In 1908 Edison entered the housing market, presumably in part to augment his cement output albeit in a somewhat ambitious form. By 1910 he had two experimental concrete buildings, a gardener's cottage and a garage at his New Jersey mansion, Glenmont.

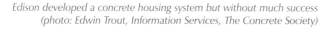

Edison developed a concrete housing system but without much success
(photo: Edwin Trout, Information Services, The Concrete Society)

A Brief History of Decorative Concrete

Edison had proclaimed in an after-dinner speech in New York that concrete homes would revolutionise American life. They would be fireproof, insect proof and easy to clean. The walls would be constructed using pigments and would never need painting. Everything from the roof tiles and balcony structure to the bathtub, picture frames and wall dado rails would be cast in one operation. His objective was to provide these for $1200 (c.£300 at the exchange rate of the time).

Early prototypes proved disastrous. The moulds were changed to nickel-plated iron and contained 2300 parts. Any builder wishing to enter the business had to pay out around £43,750 for the moulds. Edison failed to provide builders with complete plans so few attempted to build them. However, some were built by Frank Lambie and Charles Ingersoll in Union, New Jersey and are still inhabited today.

Undaunted, Edison announced he was going to make his invention freely available to anyone who wanted to try his scheme. Henry Phipps proposed to construct houses to solve the New York housing shortage problem which had been generated by immigrant workers. He suggested he could build a whole complex and rent them to families for $7.50 per month. Nothing is known of whether he succeeded or not.

What made Edison's plans even more surprising was his intention to have items such as house furniture, pianos, refrigerators and even phonographs (record players) cast in lightweight foam concrete. The whole scheme failed rapidly, although some houses do survive. Edison Portland Cement Company remained in business, lost millions of dollars and closed after his death.

Edison's company circulated a monthly brochure, The Edison Aggregate. In one edition he publicised his 'effusively hortatory promotional book entitled The Romance of Cement', the first chapter of which is entitled 'The Eternal Romance of Cement'.

This attempt at introducing concrete into mass produced housing was perhaps bound to fail, but Edison did succeed in demonstrating the possibilities of the material, including the decorative form possible. He was just too ambitious. However, more recently some of his concepts have become reality, for example decorative concrete flooring and kitchen worktops. It was to be another 50 years before concrete housing was to become a viable product and even then it was at the expense of providing a less than satisfactory aesthetic decorative appearance.

Some of the early attempts at producing a decorative finish with concrete, both Portland cement and other binder types, did succeed in illustrating the potential of these materials for providing decorative and durable building components. As with any new material, an understanding and appreciation of the fresh and hardened properties is required to successfully exploit the potential benefits.

3.6.4 Growth in the structural and decorative use of concrete during the 1900s

In 1898 François Hennebique designed and built what was claimed to be the first reinforced concrete building in Britain, Weaver's Mill in Swansea. An earlier smaller mill building was thought to have been erected in Carmarthen, Dyfed, South Wales but no trace of this remained. Now both have been removed. Hennebique and his company Hennebique

The use of 'architectural concrete' in a reconstruction of the Parthenon in Nashville, Tennessee, 1897 (photo: © Serban Enache/Dreamstime.com)

& Le Brun went on to erect a reinforced concrete bridge at Chewton Glen in the New Forest, and it is claimed that within 10 years some 40,000 structures had been erected using the Hennebique system of reinforced concrete.

This rapid growth in civil engineering applications was accompanied by a better understanding of the properties of Portland cement. G & T Earle, cement manufacturers in Hull, published their *Standard Methods of Testing Cement* in 1904. This was so well received that two further, expanded, editions appeared, and also in 1904 the first British Standard was published, BS 12: *Portland Cement*. In 1918 the British Precast Concrete Federation was formed, and white Portland cement had become available.

Through the Romans, mass concrete had established a convincing track record. It was used for the Cannington Viaduct, near Lyme Regis in Dorset, on the Axminster–Lyme Regis line. Here the object was to build a concrete structure that resembled masonry. The viaduct consisted of ten 15 metre elliptical concrete arches. The piers were built in a series of lifts about 2 metres high, each lift being reduced by about 8 cm at each end and 4 cm on each side; the dimensions of the lowest lifts of the highest pier were about 6.25 metres by 2.6 metres by 2 metres high, and in the highest pier were 12 lifts. At the top of each pier and on each side there were two rows of four corbels. Precast concrete voussoirs 0.75 metres and 0.45 metres wide, running back into the arch alternatively about 0.6 metres and 0.9 metres, were built in on the arch faces, the space between the voussoirs in the arches being shuttered concrete. The parapet wall and coping were of concrete cast *in situ*, and the decorative recesses were the result of a desire to use a simple type of shuttering and to economise in concrete. There was no decorative work except on the parapet but the overall form was appropriately proportioned so as to provide an aesthetically pleasing form akin to a natural stone structure.

During the period up to the 1940s the use of concrete in an unadorned form, for the façades of buildings, became more common. In the USA, where most of these structures were being built, this was referred to as 'architectural concrete'. No attempt was made to cover the concrete; the idea was rather to produce a surface that was acceptable as it was. A somewhat dramatic example is a reconstruction of the Parthenon erected in Nashville, Tennessee page 45. Every detail of the structure, built as an art gallery, was in either *in situ* or precast concrete, the work of one John Joseph Earley.

Earley (1881–1945) was also known in the USA for a development called 'polychrome', in which colour was introduced into concrete at the Earley Studio in Rosslyn, Virginia. A few houses incorporating this technique were built in Maryland (although more were planned) by shipping precast panels with colourful aggregate exposed on the surface. Earley died rather suddenly and written records of his casting techniques were destroyed in a fire.

Earley took over the studio from his father James Farrington Earley in 1906, and directed it together with Basil Taylor. They concentrated first on stucco and plaster, and then developed a form of exposed aggregate that was readily accepted, by stripping the forms early and rubbing the concrete with wire brushes to reveal the aggregate on the surface. His *Fountain of Time*, a monumental structure that used plaster as permeable formwork, is described in section 5.28.

The structure now regarded as perhaps Earley's most ambitious is the Baha'i Temple. Here again he practised the idea of separating the structure from the ornament, as he had done on the Parthenon in Nashville. The panels for this task became huge and thin and included units to form a large dome on the roof. Earley developed a de-watering system by sucking out free water from the cast products using burlap and paper in a repeated poultice application as the unit set. Demoulding was then possible after 18 hours. On this job he applied acid as well as wire brushing on the surface of the castings.

John Joseph Earley has been called the last of the concrete pioneers. While working on the Thomas Alva Edison Memorial he suffered a massive stroke. He left his Earley Studio to his business partner, Basil Taylor. Whereas others had perfected the mix, design and use of concrete, Earley developed a way to impart brilliant permanent colour to the surface (see Chapter 5).

In the UK and elsewhere, bulky concrete structures such as water towers were erected that exploited the ability of concrete to provide a mouldable, structural form that could be made decorative by the choice of proportion and form rather than relying on applied decoration.

By the 1930s and 1940s concrete structures with two-dimensional curves were going up, again exploiting the ability of concrete to readily take shape and form. Some notable examples are to be seen on the station building of the Southern Railway in Greater London. Some rather formidable structures were erected in the 1940s, using the structural benefits of mass concrete. An example is this observation tower (pages 50, 51) on the coastline on Jersey in the Channel Islands. It has a feeling of impregnability and permanence while at the same time demonstrating effective use of the material, collectively providing a fundamental and aesthetic appeal.

When casting against timber shuttering a bulky square structure would often be decorated in what might be called the Egyptian style, employing protruding linear ribs and rectangular repeated façade panels. Two main trends in the textural finish of concrete were evident, one striving to transfer as smooth a surface as possible, the other faithfully moulding the grain of timber. The latter was achieved by the use of unwrot strip boarding (rough sawn). An interesting further development was achieved by wedging alternate boards to bulge out from the rest. The following is from an architect involved in such a building:

> Here and there the accidental overlapping of 7/8 inch shutter boards was permitted. Where such surfaces happened to come against boards that had been kept smooth there resulted a contrast of unexpected vividness and value.
>
> In some of these surface finishes the ends of the boards are deliberately kept out of alignment, and unevenness can be further increased by using boards of different widths which the carpenter can be told to place 'accidentally'.

A board mark finish using vertical and horizontal boards in panels with 'v' joints between was in common use by architects until the end of the 1970s.

Concrete is the favoured material for casting structures such as the bridges. Here rectangular form with repeated decorative additions is suited to *in situ* or a combination of *in situ* and precast concrete, both structurally, aesthetically and for durability.

In situ concrete was also being employed to provide slender and graceful structures, especially in bridges. The bridge over Salgina Gorge at Schiers, Switzerland, 1930 is a splendid example of the period (page 51, 52). The designer was Robert Maillart, who had been designing such bridges since 1905.

Precast concrete arguably reached its zenith in the work of architect-engineer Pier Luigi Nervi, who was responsible for the roof structure of the aircraft hangars at Orbetello, Italy, 1939. Each precast unit was attached to adjacent units by overlapping reinforcement that was then encased in concrete. The resulting structure is outstanding: delicate, beautiful and innovative.

The following notes are taken from British and American practice of that period (up to 1940). They were used in the production of presentable concrete surfaces, giving consideration to both architectural treatment and practical problems of concrete production of the time.

Concrete:

- Careful proportioning of concrete mixtures and grading of aggregates [is necessary] to avoid surface honeycombing and pitting.
- Patching of holes is never satisfactory and can only be hidden by cement-washing or painting the entire surface in order to get a uniform colour.
- A cement-rich concrete gives a better finish.
- Excess water will tend to result in surface pitting: a dry concrete will tend to produce a coarse surface.
- A wet concrete will dry out lighter.

Shuttering:

- Two approaches: Smooth flat surface or timber board strip grain finishes.
- Smooth cast against plywood, steel or special boards.
- Thin mould oil application to avoid surface pitting.
- Joints can be filled with clay, plaster, mastic and a mixture of equal parts of tallow and cement.
- Any joint lines should be arranged to form a symmetrical pattern, with horizontal joints aligning

[Above and opposite] *German-built observation tower, Jersey, Channel Islands, 1941 (photos: Ross Nieuwburg)*

[Far right] *Bridge over the Salgina Gorge, Schiers, Switzerland, 1930. Robert Maillart designed curved concrete bridges from 1905 (photo: Brian Duguid)*

A Brief History of Decorative Concrete

Bridge over the Salgina Gorge, Schiers, Switzerland, 1930 (photo: Brian Duguid)

with features of the building, such as window heads and sills and door hoods; vertical lines should be related to door and window openings or other features.

- Plywood with the grain running horizontally can retain water bubbles on the surface of vertical shutters. Plywood should be painted to avoid water absorption and grain rising when it gets wet.
- Sharp arrises to be avoided by including a chamfer in the corner of the shutter.
- Panelling can be used to break up large surface areas at small cost. A contrast is formed if the panels are rubbed down and the margin bush-hammered. Reeded or fluted panels may be used plus a range of inserts that can be chosen to provide specific textures.

Finishing:

- Blemishes are not normally noticed in rough timber finished concrete.
- If a perfectly uniform colour is required it is necessary to rub down with carborundum or paint the surface.
- Architectural concrete uses decoration sparsely, flat surfaces and straight lines being relied upon for architectural effect.
- Decorative features may be cast in or precast and either placed in the shutters or fixed when the shutters have been struck.

- Decorative features can be precast using plaster moulds allowing many repeated inserts to be used. The pattern is usually designed to flow across adjacent inserts. Plaster requires a shellac coating before use.

3.6.5 Cast stone development during the 1900s

Cast stone is denoted by a wide variety of names. These include:

- reconstituted stone
- reconstructed stone
- artificial stone, often shortened to art stone
- imitation stone
- synthetic stone.

Care needs to be exercised in using the most appropriate title for the product. 'Cast stone' and

Reconstructed Portland Stone, India House, London (photo: Phil Henson, http://www.flickr.com/photos/24583241@N04)

[Above and opposite] *Reconstructed Portland Stone, India House, London (photos: Phil Henson)*

'reconstructed stone' are usually the most appropriate. Artefacts made to resemble the stone used as the aggregate in the casting should be referred to as 'reconstructed stone'. This avoids misunderstanding if the product is not manufactured using stone aggregate of the type it is trying to simulate.

'Artificial stone' is still used when a product is cast to resemble Portland stone. This is because the Portland cement binder used to cast the product also provides the pigmentation and the product only imitates one stone, namely Portland. Cast stone developed into a major industry during the 1900s, largely because the materials proved to be more resistant than many natural stones when exposed to the polluted urban environments of the 20th century.

With the inclusion of reinforcement, cast stone could be used to produce architectural features such as cornices with wide overhangs which would not have been possible with natural stone. Cast stone is cheaper than natural stone, especially when moulds are reused.

The original method of producing units had been more akin to plastering than casting. Architectural units were often

worked up to shape from cores by applying the necessary surfacing. The stone was finished by rubbing cement and stone dust into the surface mixtures and trowelling (a procedure made possible by slow-setting cements of the 19th century). This technique of applying further material to the face of a unit was not countenanced by the Cast Concrete Products Association of the early 1900s, and production changed to the use of either a wet casting concrete or a semi-dry material that required considerable compaction into and against the face of the mould.

Semi-dry material had several advantages: it provided a more realistic texture copy of natural stone; moulds could be demoulded immediately and refilled; the units could be cast with different facing and backing mixes, thus providing economical use of expensive facing mixes whilst allowing incorporation of a more structural backing concrete that could be bond to encased reinforced.

Typical applications of cast stone from the 1930s include a plaque cast using Clipsham stone aggregate, for Coventry Technical School and India House in London, showing fine balcony panels and window surrounds.

Stowe School Chapel and the Café Royal in London are examples of external elevations supplied in cast stone by the Empire Stone Co. Ltd. The latter has one facing in natural stone and an adjacent one in cast stone; both have weathered in the same manner.

Cast stone can be carved in the same way as natural stone, using fine aggregate only. Normally a clay model was prepared by the sculptor and used in a mould to cast an article that would then be finished by the sculptor. Examples of this form of decoration can be seen in a panel carved by Miss F.L. Sayers on a building on the Great West Road, London, and in the works *Hebe* and *Aesculapius* by the sculptor Gilbert Bayes.

3.6.5.1 General properties of cast stone in the 1900s
Cast stone has its advantages and disadvantages, both of which need to be appreciated in order to avoid problems in service. Cast stone, like ordinary concrete, can be moulded to form considerable projections, etc.; for such applications it has a great advantage over natural stone, with which a large part of a block might need to be cut to waste to form the necessary profile.

Reconstructed Clipsham Stone. Coventry Technical School.
A.W Hoare ARIBA (photo: Edwin Trout, Information
Services, The Concrete Society)

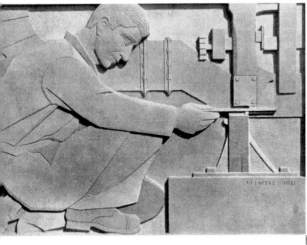

[Left] *Panel carved in concrete by Miss F.L Sayers 1932 (photo: Edwin Trout, Information Services, The Concrete Society)*

[Below] *Clay model for plaster mould by Gilbert Bayes, Sculptor (photo: Edwin Trout, Information Services, The Concrete Society)*

Decorative and Innovative use of Concrete

Cast stone has found a structural role using reinforced granite concrete backing with a limestone aggregate and white cement decorative facing mix. This is an economical compromise, with the more porous costly facing mix being a mere 25 mm thick. In addition, weight savings have been made by casting units with a hollow core.

Shrinkage

Cast stone, especially the semi-dry form, expands on taking up surface water and then shrinks during drying. The initial shrinkage produces the greatest amount of movement but cyclic wetting and drying can generate significant movement, c.1 mm in 2 metres. Thus it is often necessary to re-point joints within a year after fixing, especially for large or long units. Where coping had or has a damp-proof course beneath, hair-cracking may not require re-pointing or may not be re-pointed so as to leave a clean appearance.

Crazing

This was the most serious problem with cast stone. It was originally thought to be caused by the surface drying and shrinking whilst the bulk of the product remained damp and therefore did not shrink. The surface became stressed and when this exceeded the tensile strain capacity the surface cracked, normally in a series of closely spaced fine surface cracks (crazing). (It should be noted that ordinary concrete can and does sometimes exhibit such a crazed surface.)

It was then discovered that a second phenomenon was occurring at the surface: carbonation. The surface layer was reacting with carbon dioxide in the atmosphere, neutralising calcium hydroxide within the surface produced as a by-product of the hydration of Portland cement. The carbonation front layer was found to proceed into the concrete at a reduced rate and depth with time and as a function of the moisture content and porosity of the stone. It was determined that a typical unit might carbonate 6 mm after two or three years.

The carbonated skin had the advantage of being denser and harder, and therefore more durable, but was accompanied by shrinkage that could cause cracking. Means were then found to carbonate cast stone artificially when the material was only a few days old. The process often used flue gases from hydrocarbon-burning boilers at the plant. This factory process combined shrinkage and carbonation of the surface that was then not affected by atmospheric carbonation and so would not subsequently suffer crazing *in situ*.

It was also found that dense cast stone produced marked differences in moisture content between the surface of the stone and the material beneath. Alternate wetting and drying of the outer surface could generate crazing that might not occur in a more porous cast stone, so crazing was then associated with dense, strong products. However, it was also noted that porous material became dirty quickly and was less resistant to the pollution in urban environments.

Density

Architects of this period held the opinion that density could be equated to quality: the denser the product the better the cast stone would perform in service, with less chance of water take-up and dampness transmitted to the inside of buildings. To determine the properties of a particular cast stone, water droplets would be laid on the surface and the time taken for absorption noted. However, it was found that water penetration through masonry walls was almost always through mortar joints and hair-cracks in the joints.

Finishing cast stone

Once a unit had been demoulded, it might be finished by working the surface or carving.

Weathering of cast stone

If cast stone was to weather like natural stone, then the product needed to have a moderate water absorptivity such as found in natural stone. In towns with high atmospheric pollution a dense and resistant material was needed. Often porosity, judged by density, and price would be the factors on which products were chosen, but these alone do not give an accurate indication of weathering characteristics. The chief factors that influence weathering are:

- Surface finish: The method of finishing usually included slurrying the surface with a mixture of stone dust and cement – often on the base of poor-quality stone. This promoted crazing.
- Porosity: A moderate amount of porosity can reduce the risk of crazing.
- Maturity: Emphasis was placed on the time that should be allowed between ordering cast stone and

taking delivery. It was important this should be sufficient to allow the cast stone to harden properly so that it was not fixed whilst in the 'green' or immature condition – having to experience the full intensity of the weather before achieving a durable facing.

3.6.5.2 Manufacture of cast stone

As mentioned above, the original technique of manufacture was akin to plastering over a shaped core. The stone was finished by rubbing cement and stone dust into the surface. At that time cements were slower to set and work rates might have been more leisurely. Manufacturing of this type was carried out overseas in warm climates such as India without producing inferior materials. It was thought that dry climates had much to do with creating concrete products free from crazing defects.

Cast stone practice in the 1930s was to ensure that nothing was added to the surface of the product, and any 'skin' formed during moulding would be removed by 'dragging', tooling or grinding. A degree of surface filling was permitted (just 5%, according to the Cast Concrete Products Association) when cavities had been revealed during manufacture.

Consistence varied between 'pourable' mixes and mixes so dry and non-plastic that vibration hammer compaction was needed. Semi-dry mixes were essential to simulate natural stone. Portland cement was used in all mixes but sometimes rapid-hardening cement was employed in backing mixes because the cement was a dark colour. It was found that some brands of rapid-hardening cement dried to a whiter colour and these began to be used in facing mixes as well. Pigments were needed if the natural sand and aggregate failed to provide sufficient pigmentation to the concrete.

Aggregates were chosen for backing and facing mixes. For backing mixes normal structural aggregates, typically granite, crushed Carboniferous limestone and general ballast, were used. Facing mixes were cast using a great variety of natural stones and sands. Often when a particular stone was to be simulated, aggregates would be selected using that stone. However, this was not always straightforward. For example, if Bath stone were to be simulated, crushed Bath stone was found to produce an inordinate proportion of fine grades and dust; the dust, being soft and absorptive, made the facing very weak and liable to excessive shrinkage. This effect was corrected by sieving out material passing 3 mm (1/8 inch) and then adding another sand to replace the fines. Alternatively, another stone such as Portland stone would be used and the desired colour achieved by pigmentation.

Grading proportions were chosen to relate to the type of product being cast. Very fine detailed work was cast using aggregate all 3 mm, and smaller but deep-textured cast stone would be manufactured using a wet mix that might have 10 mm-down, graded aggregate.

Mix proportions were usually in the range:

- 1 part of Portland cement
- 3.5 parts (by volume) of aggregate
- 0–3% of pigment (by weight)

and

- 1 part of Portland cement
- 4.5 parts (by volume) of aggregate
- 0–3% of pigment (by weight).

Where aggregates such as Bath stone were used the proportions were:

- 1 part of Portland cement
- 1.5 parts of clean, washed sand passing a No. 16 sieve
- 2.5 parts of crushed Bath stone passing a ¼-inch sieve and retained on a No. 16 sieve
- 1–3% of pigment (depending on colour required).

White Portland cement, white sand, and crushed Derbyshire spar, passing a 3/16-inch sieve, without pigment, would be suitable for a Portland stone product.

3.6.5.3 Moulding cast stone

Wooden moulds

Wood was, and is, the most commonly used material for cast stone moulds because it is both cheap and easy to process into moulds. Mould making requires skill and knowledge to ensure moulds survive repeated use and still provide products of acceptable quality requiring minimal finishing. They also need to be easily disassembled and reassembled.

Wooden moulds are most commonly made from pine, deal or beech but any sound timber can be used provided it is free of shakes and large knots and is not twisted. It was normal to manufacture timber moulds with wooden sections that might appear too thick; however, it was vital that moulds withstood tamping and loading when being transported around the works whilst the concrete remained plastic.

Timber moulds were sometimes submerged in water until they were saturated. This would prevent swelling when the wet concrete was poured into them. An alternative was to coat them with cellulose paint. Mould oil was used, as also was paraffin. Moulds were sometimes whitewashed or coated with two or three coats of shellac to prevent concrete bonding before they were oiled for the first use. Current practice is to use proprietary mould coatings.

Moulds required thorough and attentive cleaning between uses to ensure detail was transferred to the concrete and to prevent products sticking when demoulding. Mould making is considered a special branch of carpentry, akin to pattern making, in which a great deal of experience and expertise is necessary.

Simple but efficient means were required for clamping the components of a mould. Perhaps the simplest and most common was, and remains, timber wedges. The choice of fixing method depends on the product being cast, the consistency of the concrete and how the concrete is to be compacted.

Rarely was a simple flat product cast – often a profiled surface was required. This was often achieved by producing a mould insert using strips of timber cut and shaped appropriately. Where a drip was required, a half-round bar would be lightly tacked to the side of the mould. If a tapered detail were needed this was sometimes formed using a tapered insert, but more often an insert would be set into the side of the mould.

When a unit is demoulded, the drip-forming detail demoulds with the component being cast and can be left in place until the concrete is strong enough to allow it to be demoulded without breaking off the arris to the drip.

Sometimes it was common to cast a unit such as a curved top for a coping or edging by screeding the surface in an open-top mould using the curved sides of the mould as screeding rails. This had the double advantage of producing a relatively cheap mould and enabling the future exposed surface to be checked before the unit was left to harden.

Plaster moulds

In cases where a great amount of detail was required it was not possible, or too costly, to use a wooden mould. Ornamental work was thus regularly carried out in plaster moulds. This involved producing a master against which the plaster could be cast.

A master or model might be made from wood, plaster, wax, clay or any other suitable material fashioned into the desired shape. The surface was treated with two coats of shellac to prevent bonding with the plaster by making it impervious to water. A coating made from 3 lbs of Russian tallow stirred into 1 gallon of near-boiling paraffin was then applied hot to the master.

Plaster was then applied over the master. This was made by adding plaster to water that had been dispersed in glue-size. The glue-size acted as a retarder, allowing sufficient time to apply the plaster; the concentration was adjusted to suit the volume of plaster being mixed. The mould makers of that time used their hands and arms to stir the plaster into the water and glue-size mix. For large moulds a range of plaster/glue-size mixtures of increasing glue-size concentration were made up, providing a continuous supply of plaster with which to coat the master and build up the mould.

Depending on the complexity and degree of curvature, and certainly if circular items were being moulded, the first stage or piece-moulds would be applied in sections of c 1-inch thickness such that they could be removed from the master. Working edges were cut to remove surplus plaster. Once the whole master had been moulded with removable pieces, the abutting edges and the backs of the pieces were greased and a backing case cast over the assembly.

The backing case held the whole assembly together but had to be removable for demoulding. It therefore had to be in sections, say three for a circular unit; these were cast against moulding clay edges applied to the desired thickness and formed by cutting to line with a knife. The formed edge was then greased and the next section cast against that edge and a clay edge formed as before.

The whole assembly was then opened up and the moulding face of the inner piece-moulds given two coats of shellac. When the second coat had dried the faces could be oiled. The inner piece-moulds were then attached to the backing case and the backing assembled to form the complete mould.

The mould would normally be held in a timber frame and the concrete poured in through the bottom. Often air holes were needed to ensure 'cul-de-sac' portions of the moulding would let air out as concrete filled the mould. The life of the mould was increased greatly if it was first heated in an oven and three coats of linseed oil applied, allowing each coat to dry before the next was applied.

The whole procedure was extremely time consuming and required great skill in devising the appropriate demarcation lines of the mould components. The exercise is also messy when a plaster model is being cast using a metal profile, dragging the tool over a wooden frame to extrude a parallel fluted detail.

Fibrous plaster was used to strengthen large moulds by including canvas scrim cut into strips, soaked in plaster and laid over the mould with a 2-inch overlap with the next piece. Sometimes wooden laths were used to stiffen the moulding as well.

Illustration 1 shows how a circular mould was made in plaster for a column base or capital. The box (a) was made the required size, about 4 inches larger than the diameter of the piece. The 'strickling bar' or template made with a zinc profile was then set in the mould as shown in Plan A. The pieces (c) were inserts used to form a square base section for a column base. The square box is filled in four sections by first installing timber moulding edges (d) so that the plaster filling can be swept to profile by the strickling bar through one quarter of the mould. The mould edges are then inserted opposite so that section 2 can be cast and formed with the strickling bar. After the four edges have been shellacked, the remaining opposite sections 3 and 4 can then be formed.

Illustration 2 shows how a plaster mould for a globe is formed in two halves by sweeping a zinc template supported on a shaft over the edges of each mould box.

Decorative and Innovative use of Concrete

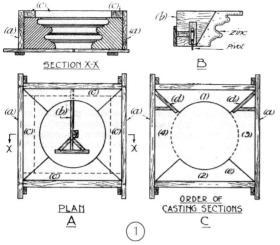

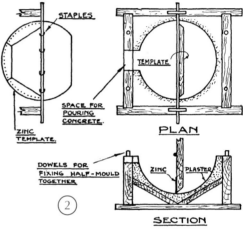

Illustrations: Edwin Trout, Information Services, The Concrete Society

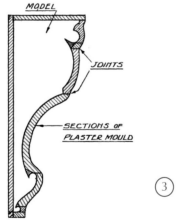

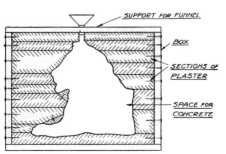

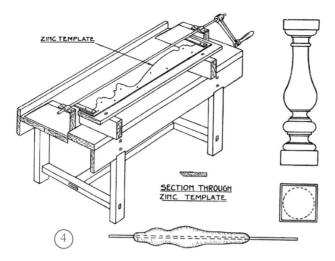

ZINC TEMPLATE

SECTION THROUGH
ZINC TEMPLATE

④

Illustration: Edwin Trout, Information Services, The Concrete Society

Illustration 3 shows how an intricate ornamental piece might be sectioned to ensure mould removal once cast, while the bottom shows how a figurine of a lion's head would have been formed in plaster slabs with the mould filled through the top.

For balusters and other round shapes a plaster model was turned in a lathe, building up plaster from the spindle to a fixed template as shown in illustration 4.

Gelatine moulds

Where much fine detail was to be moulded, gelatine moulds would be used because they were quicker to produce and easier to demould from undercuts due to the pliable nature of the gelatine. The disadvantage with gelatine was it would produce only about 10 castings before distorting during the pulling necessary on demoulding. Clay was applied to the master to form a uniform layer over the detail and profile. Plaster was then applied over the clay to form a jacket. When the plaster had hardened, the jacket was removed in sections and the clay scraped away from the plaster.

Illustration 5 shows a model with part of a plaster jacket and the intermediary clay removed. After the plaster jacket had been assembled around the figurine, gelatine would be poured between the two, forming the mould facing medium.

⑤

Illustrations: Edwin Trout, Information
Services, The Concrete Society

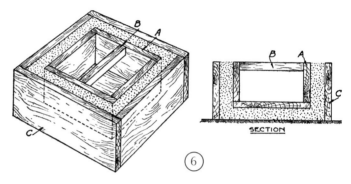

⑥

Decorative and Innovative use of Concrete

Sand moulds

Sand moulds required a new mould to be produced every
time whereas the other mould types could be reused
until either the mould was worn out or the production
run complete. The principle of sand moulding is shown in
illustration 6. The section shows an outer frame (C), roughly
constructed but strong enough to withstand compacting
the sand. A base layer of damp moulding sand, minimum 4
inches thick, was rammed into place and the wooden core
(A) placed on and hammered into the base sand layer. The
outer four sides of the mould were then filled with sand and
compacted and the core removed, leaving the mould ready
to be filled with concrete.

The reason for using such a thickness of sand is to ensure
surplus water in the wet concrete mix is absorbed by the
sand. Castings made in sand moulds have a sand face that
can be left on the product or removed by acid or other
means to leave a stone-textured face.

Illustration 7 shows concrete being placed and compacted over a curved sand mould.

Concrete moulds

Concrete moulds were, and remain, a good choice since they are easily made and relatively permanent, do not twist and can be stored in the open. They were used for items that needed to be cast occasionally over many years.

Sand was used as the aggregate to ensure a clean and accurate mould face with sharp arrises. Reinforcement was often included to reduce the bulk of concrete while still providing sufficient strength in the mould. A mix of 2.5 parts sand to 1 part Portland cement with an addition of 10% hydrated lime by weight was used to give a 'fatty' mix that was workable, copied the master and gave a good surface finish.

The method of making a concrete mould depends on the shape of the model, but the procedure was similar to that for plaster moulds except a retaining outer box was employed to contain the fluid concrete. A turned wooden model would be cut or formed in two halves along the vertical axis of the model.

The half model was then well oiled and screw-fixed through the baseboard of the mould box which would be about 2 inches wider than the widest part of the model. Shaped timber inserts were included at the narrowest part to reduce the amount of concrete required to fill the mould box and cut down on unnecessary weight. Small items would not require inserts.

Concrete was then placed around the model and into the mould box to a thickness of c.1.5–2 inches and wire mesh included if necessary to strengthen the final casting. After the concrete had hardened, the mould box was taken to pieces and the timber half model withdrawn. The process was then repeated to produce the other half of the concrete mould. The two halves would be strapped or cramped together to form the mould and alignment assisted by dowels across the mould halves.

Metal moulds

Metal moulds were preferred for high-output repeated casting. These would be bought from a specialist metal casting company and made from cast iron, steel, aluminium or other suitable metal. Sometimes they would be made

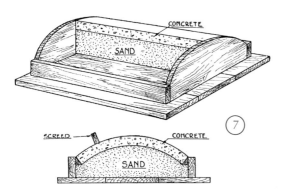

⑦

⑧

Decorative and Innovative use of Concrete

Illustrations: Edwin Trout, Information Services, The Concrete Society

⑨

⑩

from sheet steel folded and welded into shape, especially when used to cast plain masonry items.

It is evident that mould manufacture is a major and time-consuming part of the casting process. Mould production is a critical factor in determining the quality of the output, and successful production of decorative items in concrete or cast stone requires a sound knowledge of the mould-making process. Output was not limited by the moulds available at the time; in fact the items made in the 1900s were perhaps more elaborate and decorative than those produced nowadays. Tastes have changed and concrete technology has developed. Intricate decorative cast stone production is now more concerned with the concrete mix design than mould production since flexible rubber moulds can be readily moulded against masters without the need for the intricate stages in mould production outlined above.

Illustrations 8–10 show some of the kinds of decorative detail achieved. One can't help but consider the moulding complexity that was required to produce them.

3.6.6 Concrete finishes in the 1900s

Since concrete had for long been considered a grey, non-uniform and uninteresting surface, various means were developed to make the appearance more lively and aesthetically acceptable. Applied finishes and improvements to the concrete surface were developed and perfected, some involving elaborate techniques that are now no longer likely to be cost effective.

Rendered finishes

Before any surface render could be applied or considered, the concrete was thoroughly cleaned by washing down and wire brushing. Washing not only removed dirt but also dampened the surface. This prevented water being sucked out of any applied treatment which would reduce that available for hydrating the applied coating. The requirement was for a clean concrete surface with just the surface dry and the underlying concrete dampened.

A key was formed by applying either a spatterdash coat or rich cement backing coat to the walling or metal mesh fixed to lathing screwed to the wall (the choice depended on the surface being covered). Blockwork or brickwork often required a spatterdash coating to remove variation in surface

Photos: Edwin Trout, Information Services, The Concrete Society

suction that would otherwise result in the appearance in the applied render of dark lines over mortar jointing.

Spatterdash consisted of a mixture of coarse sand, cement and water in the proportions 1.5 parts sharp sand ¼-inch down to 1 part Portland cement and about 1/5 part of water by volume. The mixture was 'dashed' or thrown on the surface in an uneven manner. The spatterdash would be expected to crack as it dried but this only served to improve the mechanical lock of the render.

The render was made up of a sand/cement mixture. Horse hair was sometimes added to bond layers together, especially where mesh and lathing was used; 10 lbs of hair per cubic yard of render was the norm. Sand in the backing mix was ¼-inch down and was mixed 3 parts sand to 1 part Portland cement. Finishing coat sand would be 1/8-inch down to a ratio of 3 parts sand to 1 part Portland cement and to this would be added 8% hydrated lime by volume of cement.

At that time it was thought that cement should be spread out and left under cover for about a week before use to allow it to 'aerate'. This apparently improved the working characteristics and bond of the render.

The first or backing coat would be trowel applied to 3/8-inch thickness and after it had just set the surface would be combed with a wire-nail comb to produce a wavy lined pattern. After 30 hours a second coat would be trowel applied onto the dampened first coat.

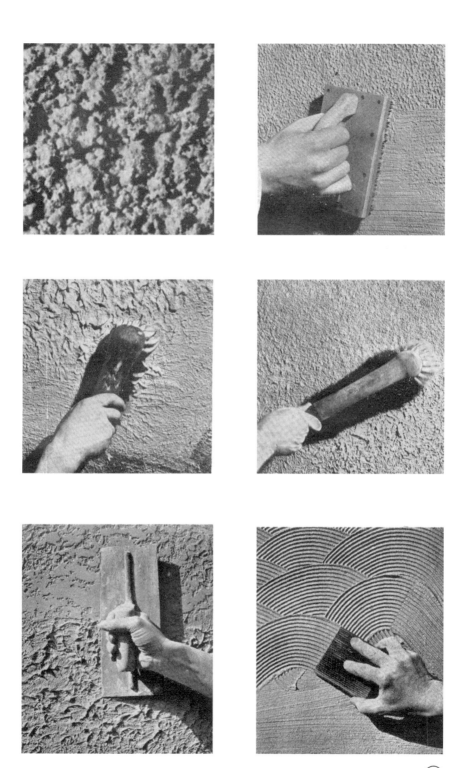

Photos: Edwin Trout, Information Services, The Concrete Society

⑫

Rendered areas had to be broken down into bays, providing joints to accommodate shrinkage cracking. Smooth render was, and is, prone to cracking. It was found appropriate to form joints to a maximum of 15 feet by 8 feet grid on walling.

One way a finish coat was applied was by using a simple machine that flicks the mixture onto the surface by simply turning a handle. Illustration 11 also shows the applied finish being scraped when it had hardened, after 24 hours, and the appearance achieved. The top left picture in illustration 12 shows the kind of texture that was obtained if left as applied by the splattering machine.

The other types of finish typically achieved at the time are shown in illustration 12 – the finishes worked whilst still plastic using a rubber brush, coarse brush, fine brush, steel trowel and using a corrugated rubber sheet. Roughcast was obtained by carrying out the application as described above but then casting an aggregate/cement mixture onto the wet finish coat, using the back of a float. The aggregate was normally in the size range $\frac{3}{8}$–$\frac{3}{4}$-inch. The finish coat used in roughcasting was usually 1.5 parts fine sand to 1 part cement.

Pebbledashing was applied in the same manner as roughcasting except the aggregate was not mixed with the finish coat of render but applied to a render that was normally $\frac{1}{2}$-inch thick. The aggregate was clean and usually washed on site.

Tooled finishes

During the 1930s a number of tooled finishes were made available. With any tooled finish it was important that the concrete was dense enough to allow surface working without voids appearing and that the surface was hard enough not to drag aggregate out of the surface. If any voids appeared these would be filled and the concrete left to harden before the work continued. If the surface texture was found to remove a voided skin leaving a dense understructure, then tooling could continue.

When tooling on a surface it was found appropriate to line the shutter face with wallboarding or place a piece of cheap muslin between the board and the concrete to prevent adhesion. The wallboard absorbed surface water and air bubbles, the chief cause of the surface pitting that can affect

*Photos: Edwin Trout,
Information Services,
The Concrete Society*

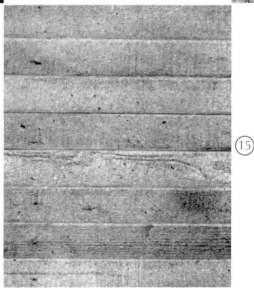

the appearance of tooled surfaces. The wallboard had to be
of the type that did not contain a waterproofing material.

Tooling concrete was the same as tooling stone and the
same skill and equipment was used as practised by stone
masons. In the illustrations a range of typical finishes that
were applied to *in situ* concrete and cast stone during the
1930s are shown.

First, illustration 13 shows the effect of wire brushing
way surface laitance to expose the underlying aggregate (top
right of image). The concrete must have a soft surface (softer

than that used for impact tooling). If the concrete is found to be too hard, bush hammering is an alternative. Bush hammering does not expose the aggregate as much as wire brushing.

The pleasing texture that can be obtained when a surface is bush-hammered is shown in illustration 14, a wall feature on Twickenham Bridge. The horizontal banding was formed by including tiles on edge at the horizontal pour joints. This has successfully disguised the joint.

The typical range of tooled finishes for both *in situ* and precast applications include:

- boasted finish
- tooled finish
- bush-hammered finish
- sparrow-pecked finish
- combed finish
- dragged finish.

Timber finishes

Rarely was a smooth, unadorned exposed finish required on concrete at that time (unlike today, where it has become the norm). Instead it was more common to texture the surface by casting against timber in various forms. Apart from giving the appearance of timber (which may or may not have been the intention) timber mould facing provided a means of camouflaging minor surface defects and could also disguise pour joint levels, for example.

Timber shuttering is relatively cheap to produce and carpenters can change and maintain moulds relatively quickly. It is best if the grain pattern is made to be noticed rather than avoided. Arranging timbers to run in different directions on adjacent panels provided a pleasing appearance. Some structures of bulky proportions benefited from intentionally offsetting adjacent boards to accentuate rather than conceal individual strips.

The ease with which concrete can take up timber grain is demonstrated by illustration 15. Variations in the depth of colour and adjacent contrasts over the moulded surface are due to variations in the absorption of the timber. Cement fines are taken to areas of the timber surface that absorb mix water and are deposited. Similarly, any grout loss through a shutter would have the same surface darkening effect where concentrations of cement occur. With repeated use timber

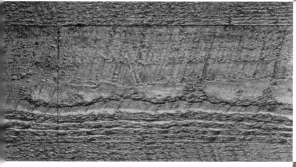

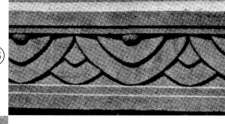

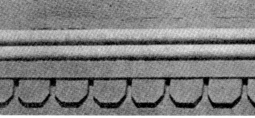

Photos: Edwin Trout, Information Services, The Concrete Society

shutter faces become less absorbent as the timber self-seals. The contrast produced by a replacement board would be clearly noticed on a shutter unless the piece was oiled or sealed before use.

Feather edge boards have been used in illustration 16 where they form 'v' joints in the face of the shutter by the inclusion of timber fillets. Both enable the joint between adjacent lifts to be disguised. Sometimes a simulation of timber is intended; illustration 17 shows how well this can be achieved.

If a decorative banding is required, this can be formed using inserts such as illustration 18 formed from plaster or illustration 19 formed from a combination of plaster and wood. Making these in timber was not as simple in the past as it is with today's profile cutting equipment.

Decorative panels on a wall are best achieved using inspected and approved precast units. However, *in situ* cast panels have been attempted; these require an experienced work team.

Achieving unusual finishes

The Dorchester Hotel has a surface which has been formed by pouring concrete on supporting concrete slabs, leaving the blocks as permanent shuttering. This has the advantage of allowing the surface to be approved or inspected before the assembly is built into position.

The finish on Marlborough College Science Building was realised using concrete which employed 19 mm Thames gravel and 6 mm sand. The cement was pigmented a light buff colour. The shutters were stripped after 48 hours and the concrete brushed after half an hour. A stiff wire brush was used, without water or acid solution. The blocks on the end of Twickenham Bridge were cast face down and scrubbed within 12 hours of casting (see illustration 20).

The abutments and wing walls of Torphin Bridge in Midlothian were faced with concrete blocks with a special finish (at least for the time it was built) produced by splitting the blocks after they had hardened.

Twickenham Bridge Cut-water and Pier were formed with a combination of two finishes. The sloping surface is smooth but the vertical surface was 'reeded' and hammered. This gave a rugged surface which was much copied later.

Decorative and Innovative use of Concrete

[Above and previous page] *Science building, Marlborough College, Wiltshire. Reeded and brushed concrete surface (photos: Peter Davies of Marlborough photo services)*

*Photos: Edwin Trout, Information
Services, The Concrete Society*

(21)

Bush hammering was used on the pylon on the Lea
Valley Viaduct, illustration 21. This technique tends to
fracture the aggregate and may alter its colour and texture.
However, the appearance of the concrete can be improved
when some of the aggregate is fractured.

Illustration 22 shows a broken-glass aggregate ceiling
at the Merchant Taylors' School. Glass aggregate has had
several popular phases over the years; this one is from the
1930s. From the same school, illustration 23 shows an
untreated internal concrete ceiling showing shutter marks,
another finish to be revived in the 1950s and 1960s.

3.7 Faircrete and William Mitchell

3.7.1 Faircrete (fibre-air-concrete)

What might be called 'normal concrete' has the property
of taking up a texture relatively easily when in the fresh
plastic state, but then slowly flattening or levelling under
the influence of gravity. Some concretes are formulated to
self-level which makes them particularly suitable for laying
floors, filling a mould in precast works or pouring into
shuttering on a construction site.

Decorative and Innovative use of Concrete

Photo: Edwin Trout, Information Services, The Concrete Society (22)

(23)   *Photo: Edwin Trout, Information Services, The Concrete Society*

During the 1960s, John Laing Research and Development Ltd developed and patented a concrete they named Faircrete (a shortened version of fibre-air-concrete). The properties of Faircrete are a result of the combination of air-entrainment and dispersion of fibres in the mix. Including discrete and uniformly spaced bubbles of air in fresh concrete modifies the consistence of the cement paste, making the concrete more fluid and easier to work and compact, as well as providing a 'fatty' surface that is readily floated to a smooth finish. Adding fibres to the mix then provides a three-dimensional sieve-like structure that overcomes segregation and bleeding, retaining the coarse aggregate throughout the concrete and preventing the air and fine particles from rising to the surface.

Without fibre in the mix, Laing claimed air content was limited to c.10% by volume, above which air bubbles can become unstable and prone to collapse, with consequential bleeding of the concrete. By combining fibre and air in judicious proportions, it was possible to entrain up to 30–40% air by volume without segregation or bleeding.

The bubble structure also had the effect of augmenting otherwise deficient sand grading and particle size distribution such that harsh sands, otherwise inappropriate for providing a good finish, could be used. Faircrete, above all, had the distinct advantage of being highly workable when subjected to small amounts of vibration but remaining static when working was stopped – a truly thixotropic behaviour.

This fresh thixotropic property was exploited and ways developed of applying texture to the surface, removing the need for casting against moulds and allowing the introduction of all kinds of intrinsic surface patterns. Texture can be applied by hand tooling but repeat work such as cladding was textured by machine, using oscillating blades with variable pitch and amplitude. This produced an almost limitless variety of patterns. The Vortex will be known to those of us who walked between the Cement and Concrete Association Training Centre and Wexham Springs, Slough before the unfortunate sale and closure of those establishments.

3.7.2 Mixing Faircrete

Practically any conventional mixer can be used to mix Faircrete. Laing found it appropriate to use coils of polypropylene monofilament cut and fed directly into the mixer, using a device they developed specially for the purpose. Polypropylene is not easily cut and is best chopped via a revolving blade. By dispensing the chopped fibre into the mix, fibre bundling can be avoided and uniform dispersion more easily obtained.

3.7.3 Faircrete fresh properties

The thixotropic property of Faircrete is such that:

- mixes can be transported without the danger of segregation
- it can be easily made to flow with only a minor amount of vibration
- it can be levelled easily by hand tamping

- holes formed in the concrete are retained after formation
- three-dimensional texturing of the surface is completely retained
- a formed pattern surface can be inspected for completeness and if necessary re-textured to ensure a satisfactory finish
- sloping retaining walls can be formed *in situ*
- reduced water absorption provides a surface that weathers more uniformly than conventional concrete
- very smooth floor and soffit surfaces can be formed in one operation without the need for screeds and rendering.

3.7.4 Faircrete hardened properties

Faircrete has a range of mix densities and strengths, which depend on the choice of aggregate. Although the cement/aggregate proportions are the same, the hardened concrete can have significant density reduction whilst maintaining similar drying shrinkage and moisture movement.

There is a significant reduction in water absorption when a mix is modified into a Faircrete version (this is more apparent with dense natural aggregate concrete). The reduction is more than accounted for by the change in density. This is better illustrated by ISAT (initial surface absorption test) performance on plain and Faircrete concretes using the same aggregates and mix proportions.

Faircrete has been found to have a good fire resistance due to less likelihood of spalling, particularly when lightweight aggregates are used. The fibre content is small – around 0.1% of the mix by weight – and has no influence on the compressive strength. Fibre appears to act as local reinforcement against shrinkage and crazing, and Faircrete is less susceptible to these phenomena.

3.7.5 Faircrete in the hands of William Mitchell

William Mitchell has an international reputation for working concrete into sculptural forms (see also page 209). He became aware of the possibilities of Faircrete and developed the idea of transferring a soft charcoal image onto the flat wet surface so he could carve and work it into a vivid and exciting relief before the concrete hardened (c.1.5 hours).

Four tablets from the Stations of the Cross by
William Mitchell (photos: William Mitchell Designs)

Decorative and Innovative use of Concrete

A series of 14 tablets depicting the Stations of the Cross were commissioned for the new Clifton Catholic Cathedral. These measured 1.5 metres × 2 metres and were produced by transferring a drawing to the smoothed face of the concrete. This image was then worked into the concrete before it set. When cured the tablets were built into alcoves, each illuminated from above.

The effect that has been achieved is splendid, combining as it does the use of a novel material with the individual and brilliant imagination of William Mitchell.

3.8 Naturbetong concrete finishes

Concrete is an ideal building material. During the 1960s it tended to be frowned upon by architects as a drab cladding material, so effort was made in Scandinavia to develop an exposed aggregate form capable of producing a consistent appearance with minimal lift lines and voids, and with surface discipline that enabled form and shape to be achieved. This is known as Naturbetong.

Naturbetong is carried out as follows:

- A rigid mould or confining shutter is assembled, grout tight with a smooth surface. Moulds need to resist deformation under the head of mortar to be injected.
- Injection pipes c.25 mm diameter are located at 750 mm centres vertically in the shutter, initially set down on the base of the shutter.
- Inspection holes are drilled through the vertical face of the shutter 600 mm vertically, 750 mm horizontally, spaced midway between the vertical tubes.
- The chosen aggregate (passing 40 mm, retained 20 mm, usually rounded) is placed in the shutter in layers under vibration to ensure uniform packing with minimum voids. Aggregate is taken to within 100 mm of the top of the mould.
- Mortar requires thorough mixing. During the 1960s the water/cement ratio was critical to ensure sufficient fluidity to allow pressure injection but not so much that segregation occurred. Modern workability admixtures assist injection.

- At the start of mortar injection, the pipes are raised about 100 mm from the base and then the outlet kept below the mortar level. It is normal practice to start at an edge and fill in horizontal bands working across a particular level up to the level of the first inspection holes which are then plugged as the mortar reaches that level.
- As mortar rises to each inspection hole level the injection pipes are raised, keeping the outlet about 100 mm below the mortar horizon.
- When mortar reaches the top of the aggregate level, kept below the shutter top edge, it is allowed to rise above the aggregate and fill the mould. The top band of mortar is then filled with aggregate by hand to ensure full compaction and a horizontal top line to the concrete.

The shutters are normally left in place for periods of up to 30 hours, governed by the temperature during curing, and any sandblasting is carried out as soon as possible after removing the shutters.

Sometimes dextrin surface retarders are applied to the shutters to assist removal of the mortar. It is possible to introduce decoration by carefully sandblasting part of the surface, forming a striated finish with clean lines of exposed aggregate.

The decoration can be taken further, as seen on Bakkehaugen Church, Oslo, where the outside walls are exposed in a pattern resembling the elevation of the building (page 89, 90). Inside the church a wall mural portraying a manger has been produced by careful exposure and localised introduction of contrasting aggregate.

In the hands of experts special sandblasting equipment can produce fine detail. Products can be cast with fine lines and shapes, although the surface mortar is removed. Curved profiles are possible which provides a surface capable of withstanding severe Norse winter weather.

The technique has also been used to produce relatively thin façade slabs by casting Naturbetong panels and jointing them with stainless steel ties to form inner and outer leaves, with the core filled with insulation.

[Above and previous page] *Bakkehaugen Church, Oslo (photos: Jan-Tore Egge)*

Decorative and Innovative use of Concrete

3.9 Artificial marbling

Marbling and other polychrome decorative effects have been obtained in precast work by arranging discrete mixtures of cement and pigment beside each other without intermixing on the mould face. Often cement and pigment are sieved to ensure thorough mixing.

The mix forming the base colour of the marble simulation is placed over the moulding table and various pigmented mixtures intermingled with the base mix. This generates veins of intermingled colour through the base colour. Sometimes the consistency is altered such that the mix resembles moulding clay which can be kneaded and/or rolled. Lumps are then broken off the mass and dipped into a pigmented vein mixture which has the consistency of cream. The lumps are kneaded back into the mass and formed into one piece, which is then is left for about 15–20 minutes.

Slices of the required thickness can then be cut off and pressed against the base of a mould or trowelled onto an *in situ* structure. For clearly defined veins the base mixture has to be of a sufficient stiffness; a dry vein mixture is then sifted over the surface and the whole kneaded together. If broad veins are required a syrupy mixture is best.

Several repetitions of the same process are carried out until the desired appearance is obtained, which depends greatly on the experience of the operator.

The best tools for placing fresh marble mixture into moulds or on walls are brass trowels, which are much easier to use than steel ones. Layers are placed a little thicker than required and then trowelled into shape. This produces a smudgy appearance, often without any veins showing. Only when the outer 2 mm is removed do the veins become visible.

Marbling can be imitated by soaking pieces of cord in a strong suspension of a pigment. The cords are laid against the forms in a random manner, the base mix added and the cords slowly pulled out. The base mix flows in to take the place of the cord and absorbs some of the pigment, forming a vein. This can be used on both horizontal and vertical surfaces; it can provide a uniform appearance and has a relatively quick casting time.

When a vertical *in situ* area is to be overlaid, marbling mixture can be laid on canvas and then pressed against the surface. This is then trowelled over to obtain a uniform thickness, after which the canvas is removed. A galvanised mesh is often attached to the surface and the mixture overlaid into the mesh; this ensures a key to the surface.

Before the marbling hardens, the surface is usually rubbed over with a rule coated with chalk to highlight areas of unevenness, which are then rubbed off with a scraper, working in all directions. Any small air holes are also filled at this stage.

If the surface is to be ground, grinding and polishing stones are often cast for the purpose. Grinding stones can be made as follows:

First grinding	Mix 2 parts very fine quartz sand with 1 part Portland cement. Add sufficient water to make a semi-wet mixture then tamp into a suitable mould. After 24 hours hardening, cure under water for six weeks and then in air for six weeks.
Second grinding	Mix 1 part balsam rosin with 1 part emery powder. Melt shellac in an iron vessel and add the mixture until a thick syrup consistency is obtained. If too stiff, add yellow glycerine to allow the mixture to be poured into a mould.
Polishing stones	Mix 1 part emery powder with 1 part flowers of sulphur and 1 part tin ashes. Melt shellac and combine as above. If slabs are to be cast (for example as kitchen work surfaces) the marbling mixture can be pressed onto plate glass. Very little polishing will then be required. Mould sides can also be lined with glass. Veining can be achieved with a glass mould by introducing cream-consistency pigmented

cement slurry using a teaspoon held at 45°, followed by the base marbling mix.

After demoulding, it is common to work a solution of 12 volumes of lime water dispersed in 1 volume of potassium silicate into the surface to act as a hardening, polishing fluid.

3.10 A few leading exponents of concrete in the 1900s

One of the objectives of this book is to demonstrate various ways in which concrete can be made 'decorative'. It is not intended to provide a structured historical account of architectural developments during the 1900s. However, since from around 1900 onwards concrete became the predominant construction material used by the Modernist architectural movement in Europe and the USA, it is considered worthwhile providing some account of that period. This is achieved through highlighting some buildings by significant architects/engineers of that period who by their examples brought concrete into regular use. The examples are necessarily selective, and interested readers are encouraged to explore other works by these architects and their peers from the period 1900–1960.

The following examples all come under the generic name 'Modernism'. This movement set out to emphasise function – providing for specific needs without the use of elaborate decorative embellishment. Design and form were the decoration. Overlapping with Modernism was the International style (also called Formalism). This grew out of the disbanded Bauhaus Institute when members moved in numbers from Germany to the USA, where the style of flat roofs, smooth elevations and box-like structural form continued.

Another development from the Bauhaus movement in America was the Art Moderne style. Here rectangular, box-like forms gave way to curves, triangles and cones but still with no sign of ornamentation. This was the 1930s style reflected in the building for the 1933 World Fair in Chicago, and gave rise to simple designs for houses that were easy and economical to build.

Concrete was the perfect material for this era. The elaborate decorative styles were out and the sleek energetic machine age was in, reflected in the buildings of the time.

3.10.1 Auguste Perret (1874–1954)

Perret was born in Belgium and became a leader and specialist in the use of reinforced concrete in construction. He blended modern theories with Gothic forms and showed concern for detail and texture. He is considered an initiator of the Modernist movement, in the way he connected natural forms, classical symmetry and order with the structural form of concrete, and is thus seen as the father of the Early Modern style.

Although he considered concrete as superior to masonry he viewed each element separately. He did not use it to form a structural whole, in the way later adopted by Le Corbusier. Early work, for example the Apartment Building, Rue Franklin, Paris, 1903, employed a Hennebique reinforced concrete frame with an interior that anticipated Le Corbusier's later development of a free open plan arrangement; the apartment walls were non-structural throughout. On the façade were decorative panels separated from the reinforced concrete frame. The decorated panels were simply for decoration – not a full Modernist outlook.

His later building Church of Notre-Dame, Le Raincy, Seine-Saint-Denis, 1922–23 is considered the first architecturally satisfactory building in reinforced concrete – tall with thin columns supporting low arching vaults and a continuous wall of glass supported by prefabricated concrete units.

After the Second World War, Perret contributed plans for the rebuilding of Le Havre. He was one of the most respected French architects of his generation and a prime mover in the use of concrete.

3.10.2 Frank Lloyd Wright (1867–1959)

Wright's houses were laid out with few dividing walls and rooms flowed from one to another. His open plan designs were popular and used in large houses for the wealthy as well as in flats and smaller homes for the middle class in Europe and other places.

Without doubt he is one of the most prominent and influential architects of the first half of the 20th century. Wright enjoyed living a highly stressed life which generated many problems to be solved. His personal life makes interesting reading.

Decorative and Innovative use of Concrete

[Above] *Detail of exterior cladding, Rue Franklin, Paris, 1903 (photo: Nathalie Tison)*

[Left] *Church of Notre-Dame, Le Raincy, near Paris, 1922–23 (photo: Julien Gouiric www.flickr.com/darkcorners)*

[Below] *Interior of Unity Temple considered by some as one of Frank Lloyd Wright's highest achievements (photo: Angela Abbott. Copyright 2009. Angela Abbott. All Rights Reserved)*

A Brief History of Decorative Concrete

Decorative and Innovative use of Concrete

[Above and lower left opposite] *Interior of Unity Temple (photos: Angela Abbott. Copyright 2009. Angela Abbott. All Rights Reserved)*

[Upper left opposite] *Unity Temple, Oak Park, Illinois, 1904 (photo: Angela Abbott. Copyright 2009. Angela Abbott. All Rights Reserved)*

A Brief History of Decorative Concrete

[Above] *Fallingwater, Mill Run, Pennsylvania, 1935 (photo: Pete Kreps)*

[Opposite] *Solomon R. Guggenheim Museum, New York (photo: Phil Henson, http://www.flickr.com/ photos/24583241@N04)*

Decorative and Innovative use of Concrete

Fallingwater, Mill Run, Pennsylvania, 1935 (photo: Pete Kreps)

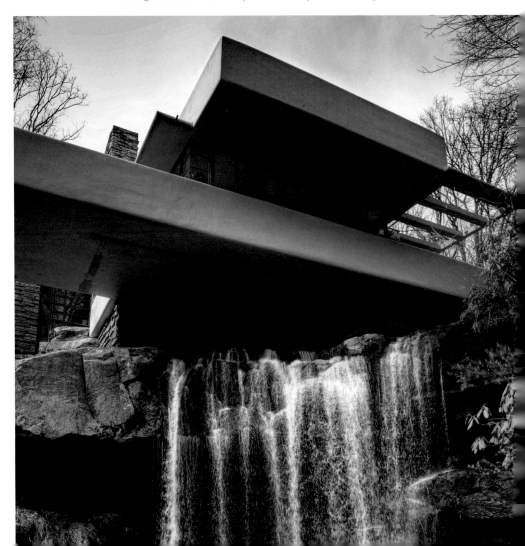

By 1901 he had completed c.50 projects, including many houses. His first public building was the Unity Church, Temple and House, Oak Park, Illinois, 1904. This was the first public building of any type in America to be built entirely of exposed concrete. Construction costs were kept low but the building was built in character.

The Temple is in the form of a cube with the ceiling opening into a grid of beams between which are set 25 stained glass skylights. Clerestories run across the top of each balcony under the roof that fills the building with light from above to provide colour and patterns from the leaded windows. Wright stated the space is flooded 'with light from above to get a sense of a happy cloudless day into the room ... the light would rain or shine, have the warmth of sunlight'. Most critics consider this one of Wright's highest achievements.

One of Wright's last buildings was the Solomon R. Guggenheim Museum, Manhattan, New York, commissioned in 1943 and completed after his death in 1959. Construction began in 1956 after the building moratorium imposed by the Second World War. The spiral form went through several changes in design, either flaring out from the bottom or reducing towards the top.

The primary construction material is concrete, sprayed and poured *in situ*. The inside follows the curvilinear profile of the outside, and the building is open and airy. Wright described it as

> one great space on a single continuous floor. The eye encounters no abrupt change, but is gently led and treated as if at the edge of a shore watching an unbreaking wave ... one floor flowing into another instead of the usual superimposition of stratified layers. The whole is cast in concrete, more an eggshell in form than a crisscross brick structure.

Two daring and now iconic works utilising concrete were built by Wright in the 1930s. The first was the Johnson Wax Headquarters, Racine, Wisconsin, 1936 (page 6). A new headquarters was needed for Johnson Wax and Wright proposed a novel column design that would provide an open plan building – his 'upside-down column' as it was referred to by the press at the time. The column defied so many building laws that a load test was ordered before it was passed for construction:

> The new column takes the form of a flower, or ice cream cone. Wright prefers to call it a 'flower' column. At the ground, where most columns have their greatest thickness, the Wright column is nine inches (23 cm) in diameter. Instead of tapering, it spreads gradually, like the stem of a flower. At the top of the 'stem', simulating the botanical construction of a flower, there is a perceptible widening of the forms to create the appearance of a cup. Wright calls this the 'calyx' from the botanical name for the corresponding part of a flower.
>
> Surmounting the 'calyx' is a large concrete 'dish', 8.5 feet (2.6 m) in diameter, which is called the 'petal'. The roof of the building will rest on these concrete 'petals', spaced 20 feet (6.1 m) apart throughout the building. Light will be brought into the building through glass skylights which will fill the diamond areas on the roof caused by the rims of the petals. [Extract from *The Millwaukee Journal*, June 4th 1937, "Wright's Upside-Down Column Tips Over Theories" "Holds heavy load in test"]

The column was required to carry a test load of 12 tons; it eventually failed at 60 tons. At that time a 9-inch diameter base was considered suitable to support a maximum column height of 6 feet 9 inches (2.05 metres). The 9-inch column supports a roof height of 21 feet 7½ inches (6.6 metres).

According to Wright, the secret in the column design was the use of a reinforcing mesh rather than reinforcing bars. He claimed the mesh provided 'muscles and sinews' rather than the 'bones' of steel bars. The columns were cream coloured with matching mortar.

Pyrex glass tubing was used for the roof glazing; this eventually required rubber gaskets to prevent water ingress. But just look at the splendid open feel the building provides for staff and the amusing way scaled down column shafts were used for the undercourt while retaining the same column heads and calyx diameter; concrete design at its most impressive. The entrance showed another adaptation of the columns.

Wright was nearly 70 when this building was erected.

The second example is his most famous private residence, built for Mr and Mrs E.J. Kaufmann and listed

Decorative and Innovative use of Concrete

by Wright as the Kaufmann Residence, Fallingwater, Bear Run, Pennsylvania, 1935. The design of this building put the occupants of the house as close to the surroundings as possible. A stream runs under the house and over a waterfall (page 100). A sketch and section drawing detail how Wright planned the layout using a series of balconies and walkways constructed in concrete, with breathtaking projections over the flowing waters. There was some concern that the design lacked sufficient reinforcement, and extra steel was secretly added. A recent refurbishment in 2002 added further post-tensioning to the lowest terrace, but the structure was then *c*.70 years old.

The way the structure was built around the site can be seen by a rock penetrating the floor of the lounge, in front of the fire. Plain concrete balconies seem to hover over the landscape and slot sensitively into the rocks in a way that preserves and augments the site rather than destroying its natural beauty (page 99). The concrete balconies were originally to be covered in gold leaf to mimic the colour of the falling leaves in autumn, but Kaufmann found this rather extravagant and they were painted beige. The balconies look as if they are about to fall but are anchored through the stonework of the hillside. The heaviest portion of the house is set at the back.

A path ascends to a guest lodge and rain protection for the walkway has been formed from concrete plates, reflecting the steps of the path in a circular sweep up the hill. The innovation is in the thinness of the plates and curved line that has allowed just single column supports to the outer curve.

The vertical elements, stairs and chimney of the building are in stone from a local quarry and the horizontal balconies cast in plain smooth concrete. So often plain smooth concrete is given a bad name and considered inappropriate; but in this context Wright has cleverly used the most appropriate form of concrete to make the interior of the house continuous with the outdoors, linking the house and occupants to the location.

3.10.3 Rudolf Schindler (1887–1953)

Rudolf Schindler was an Austrian-American who worked in Los Angeles. He trained and worked with some of the foremost practitioners of the Modern movement, such as Frank Lloyd Wright in his Taliesin and Chicago offices. However, his original use of complex designs and practice have placed him in the list of great architects of the Modern movement in his own right, although acclaim came mainly after his death.

Schindler's early work centred on the Modernist style. This was progressive for the time and may account for him working only in California. His Kings Road House, 1922, was the first to respond to the unique climate of California and was intended as a prototype for a California style. He lived in it from 1922 until his death in 1953.

The floor plan took the form of several 'L' shapes and the construction featured tilt-up concrete panels cast on site. In a bid to achieve more economical construction, Schindler later abandoned concrete and developed his plaster skin design on a frame system. He also developed a 'theory of proportions', two years before that published by Le Corbusier.

However, throughout the 1920s Schindler continued to experiment with concrete. After using tilt-slab construction in the Kings Road House he turned to poured concrete in the 12 units of the El Pueblo Ribera Courts, La Jolla, San Diego, California, 1923. Walling for one unit served as a garden enclosure for the next. The formwork was simple and ingenious. There is apparently a spectacular view of the Windansea Beach from where the sand was taken to cast the concrete.

In his Lovell Beach House, Los Angeles, California, 1926, the living space is formed inside five free-standing concrete frames cast as two squares on top of each other to form a figure of eight. The reason for such an elevated house was to raise the property above the sight line of the public beach. The frames were also intended to secure the structure in case of an earthquake.

The work of Schindler is closer to the International style than that of Frank Lloyd Wright.

3.10.4 Le Corbusier (1887–1965)

A well-recognised name in the UK, Le Corbusier was born as Charles-Edouard Jeanneret in La Chaux-de-fonds, a little town in northwest Switzerland. In 1907 he moved to Paris and worked in the office of Auguste Perret, and then went to work with Peter Behrens near Berlin.

[Above] *Villa Savoye, Poissy, France,*
1929–31 by Le Corbusier
(photo: Marcel Frei)

[Right] *Concrete stairs, Villa*
Savoye (photo: Marcel Frei)

[Top of page] *Schindler's Lovell*
Beach House, California, 1926
(photo: Ilpo Koskinen)

Le Corbusier adopted the name as an altered form of his maternal grandfather's name 'Lecorbésier', in the belief that anyone could reinvent himself. Some historians say the name translates to 'the crow-like one'. He established an art movement known as Purism. He came to know Ludwig Mies van der Rohe and they, together with others, are famous for what is now called Modernism or the International style of architecture.

At this time he began theoretical studies using modern materials, namely 'concrete'. One project, Dom-ino House, 1914–15, consisted of concrete slabs supported by reinforced concrete columns around the edges, with a stairway to each level that presumably stiffened the structure. This system became his area of focus for the next 10 years. He then took up his own practice with his cousin Pierre Jeanneret.

Various stages of design advanced until Villa Savoye, 1929–31. This became a classic Le Corbusier design and was used by him to describe his five points of architecture in his book *Vers une Architecture*, developed through the 1920s. The illustrations of Villa Savoye demonstrate these five points:

1   Le Corbusier lifted the building off the ground onto columns he called 'pilotis'.
2   The pilotis provided the structural frame that gave him freedom to design the outer façade walls as he wished.
3   The structural frame also meant non-supporting internal walls, giving the designer freedom of location and form.
4   Long strips of window could be included, providing an open view of the surroundings.
5   A ramp could be provided rising from the ground level to the roof level/floor, permitting a continuous walkway through the building.

By now the observant reader will have gleaned the interplay between the preceding architects and how thinking was becoming directed to the Modernist or International style.

Concrete was the material for this era. Le Corbusier is perhaps best known for the Church Notre Dame du Haut

Notre Dame du Haut, Ronchamp, France 1950–54
(photo: © Mihai-bogdan Lazar/Dreamstime.com)

or Ronchamp, 1950–54, a reinforced concrete building on a hillside, cast with walls 1.2–3.6 metres thick. Here Surrealism is the key; Modernism has moved on a league. We have an oblong nave, two side entrances, an axial main alter, three chapels beneath the tower, and rough walls faced with gunite (a sprayed form of concrete) and whitewashed; the roof is a contrasting concrete chunk.

Like Schindler, Le Corbusier produced a theory of proportions based on subdividing the human body frame into elements that provided ratios for horizontal and vertical scaling.

3.10.5 Berthold Lubetkin (1901–90)

Lubetkin, a Russian émigré, pioneered the Modernist design movement in Britain with his architectural colleagues in his practice Tecton and his working collaboration with the civil engineer Ove Arup.

In the 1970s Lubetkin described his life as three phases: 'Born into one world, tested in another and abandoned in a third'. The first phase was his early life in Russia before the Revolution; the second the evolution of the Modern movement and his involvement in the UK; and the last his abandonment by post-Second World War conservative Britain and his early retirement on a farm in Gloucester.

Lubetkin had been influenced by the work of Le Corbusier during a spell in France. He arrived in Britain in 1931 to work on a commission for a house for the Harari family. Concrete was only slowly becoming used in Britain; large structures were sticking faithfully to traditional construction materials but small buildings such as houses, garages, shops and factories were beginning to use reinforced concrete. Within a year Lubetkin formed Tecton and the practice undertook a commission to design a Gorilla House at London Zoo, 1932–33. This might at first seem an unenviable task but it allowed Tecton to exploit the possibilities of concrete to mould the building into a circular form, using sliding screens to vary the environment for the animals. It was at this time Lubetkin contacted Ove Arup, a Danish civil engineer working for Christiani & Nielsen in England. The building benefited hugely from Arup's brilliance and this was the start of a lifelong collaboration between the two men.

Decorative and Innovative use of Concrete

[Above] *Highpoint One, Highgate, London, 1933–35 (photo: Phil Henson, http://www.flickr.com/photos/24583241@N04)*

[Right and opposite page] *Dudley Zoo, West Midlands, 1935–7 (photos: Author's collection)*

[Previous page] *The Penguin Pool, London Zoo, 1933–34 (photo: Bill Bowdish)*

In 1934 Tecton was given the chance to design a new Penguin Pool at the same zoo. Here they contrived a classic design in the Constructivist form, a double helix in concrete winding down to the water from opposite sides in two half-circle walkways. Lubetkin and Arup devised a staggering, apparently gravity-defying, thin concrete construction in the middle of the pool. This provides a brilliant practical feature for the penguins and is an excellent demonstration of an other-than-utilitarian use of concrete.

Tecton were then invited to design two further zoos, at Whipsnade in Bedfordshire and Dudley (now within the West Midlands). At Whipsnade Lubetkin also built two small villas. At Dudley, 1935–37, Lubetkin was given a fresh site on which to introduce the public to Modernism and provided a unique collection of artefacts in that singular style laid out around the Castle.

I live a short walk from the entrance and pass by virtually daily, and I greatly lament the neglect of such a distinct collection of buildings. One has been demolished and another altered beyond recognition as the Moat Café.

Tecton and Ove Arup were allowed to explore their interest in form and shape, constructing concrete enclosures to fit onto the sloping wooded site. The *News Chronicle* reported:

> The new Zoo would be at once a scientific centre and an example of an ultra-modern town plan.

The 13 original pavilions were set in the sloping grounds around the Castle mound, though a further 60 unrelated intrusions detract from the original clean and organised layout. The future for this collection of early concrete buildings is not looking good, despite several attempts at plans to develop the site.

Reinforced concrete was the only building material that could have coped with the terrain, as it allowed detailed decisions to be made on site as to levels and whether to raise a structure up on pilotis or mould it to fit the contours of the site.

On the opening day the zoo was besieged by a flood of visitors. Buses and coaches brought 75,000, trains 60,000, trams 40,000 and private cars 80,000, all into Dudley. After 50,000 had been admitted, 200,000 were left to roam the town.

At the same time as Dudley Zoo, Tecton was designing Highpoint One, Highgate, London, 1933–35 (page 110) and then Highpoint Two, 1937–38. These were developed together with Ove Arup, and used monolithic and slab reinforced concrete. They also introduced climbing formwork shuttering that until then had been unused in housing. Lubetkin lived in Highpoint Two.

After the Second World War Finsbury Council asked Tecton to resume designing a series of public housing schemes, including Spa Green Estate, 1943–50. A decorative concrete panel over an entrance depicts the typical family unit.

Lubetkin's hopes for the post-war era were cruelly shattered by the conservative attitudes towards architecture. He resigned from a project that might have been his architectural pinnacle – designing a residential new town at Peterlee – and after completing a few developments in London effectively retired into exile in a farmhouse in

Decorative and Innovative use of Concrete

Gloucestershire, bitter at the thought of being abandoned, until his death in 1990.

3.10.6 Ove Arup (1895–1988)

Ove Arup was an engineer who specialised in reinforced concrete structures. He was born in Newcastle but studied in Denmark and began working there for Christiani & Nielsen, moving to London with the company in 1924. In 1933 he joined J.L. Kier & Co. as chief designer and director, and in 1938 set up in partnership with a cousin. In 1946 he formed Ove Arup & Partners, which has grown into the Arup Associates we know today.

Ove Arup was committed to the development of reinforced concrete and worked alongside Tecton on various projects (see previous section). One of his early Christiani & Nielsen projects was Labworth Café, Western Parade, Canvey Island, Essex, 1931–32, which was part of the sea defences contract undertaken by the company. The circular central tower is supported by a ring beam and column main load-bearing structure. Large windows overlook the Thames estuary and the building has a nautical look, resembling the wheelhouse of a ship. Arup described the building thus:

> Architecture on the cheap by an amateur architect employed by a contractor, and a client with no money to spend.

He later recalled that he was only allowed to visit the site once.

The extraordinary domes of the Dunlop Semtex Factory, Brynmawr, Blaenau Gwent, South Wales, 1947–53 were described as a masterpiece of the modern age, yet in 2001 they were demolished. The factory was designed by members of the Architects Co-operative Partnership (now operating as ACP). All were also members of the MARS (Modern Architecture Re-Search) group. Ove Arup was then the main consulting civil engineer but his contribution was not limited to engineering aspects. The scheme was the brainchild of Lord James Forrester, whose aim was to bring large-scale post-war work to the South Wales valleys, ravaged by the Depression in the 1930s.

Nine domes covered the central production area. These innovative reinforced concrete structures represented a

A Brief History of Decorative Concrete

[Above] *Labworth Café, Canvey Island, Essex, 1931–2 (photo: Dave Bullock)*

[Right] *Nervi's Palace of Labour, Turin (photo: Katerina Gordon)*

[Opposite] *The Small Sports Palace, Rome, 1955 (photos: Antonio Cibelli)*

114

Decorative and Innovative use of Concrete

construction method first developed in Germany, and provided a light and spacious working environment. The building was recognised as the most inventive industrial structure of its time, and was applauded by architects such as Frank Lloyd Wright and Le Corbusier, both of whom visited the site. The building closed in 1982 and was listed in 1986. Despite local opposition and calls to save it, the factory was demolished in 2001.

3.10.7 Pier Luigi Nervi (1891–1979)

Nervi, an Italian engineer, was born in Sondrio and obtained his degree in Bologna in 1913. He was also considered an architect but his steel and concrete designs took precedence in the 1930s due to the boom in construction projects at that time.

Applying intuition in design was as relevant to Nervi as pure mathematical considerations; for example, employing rib profiles and vaulting in large span roof designs to reduce the need for supporting columns. His philosophy was to

A Brief History of Decorative Concrete

The Small Sports Palace, Rome, 1955. Precast ribs
requiring only joint filling and the whole whitewashed

use shape and form combined with skilful fabrication to minimise material weight, and it was through this that he arrived at his splendid architectural structures.

Nervi used concrete in three forms: cast in place (*in situ*), precast and ferro-cemento (the last more akin to steel wire wrapped in a thin concrete coating). According to his published accounts, he appeared particularly fond of using reinforced concrete in his designs:

> The relationship between aesthetics and technology in building has acquired a new richness and variety with the introduction of reinforced concrete, the most fertile, ductile and complete construction process that mankind has yet found … artificial 'superstone'. Two fundamental characteristics differentiate it from all other building materials ever used by mankind: (1) the fact that it is produced in a semi-fluid state and therefore can be molded into any form, and (2) its static unity which, because of the monolithic nature of construction in reinforced concrete, holds together the various parts of the structure.

Cast in place work

Nervi's first major work was the Municipal Stadium of Florence, 1928, which seated 35,000. By this time Nervi was co-owner in a design and build company. In 1936 he carried out the construction of two airplane hangars for the Italian Air Force. Representing a competition entry and following a tight budget, these were said to be the most difficult of his working life:

> After trying several solutions based on traditional schemes (a large rectangular beam placed on the side of the doors sustaining a series of secondary beams spanning the width of 100 m), none of which seemed to me either economical or suited to the construction process, I turned to a unified structural system which found its fundamental static substance in a vaulted form. In other words, this was no longer a combination of trusses and beams, but a single resisting organism where the stresses due to its own weight and the external forces of wind and snow could be spontaneously diffused until they reached the supports placed on the three sides and the central column on the side of the doors.

The asymmetrical design, with the large open doors on one side generating a lack of support, required flexural stiffness to be established. Nervi chose a double series of ribs (1 metre deep, varying from 12 to 20 cm wide) with a horizontal truss across the door. The design was verified at the Polytechnic Institute of Milan using experimental modelling.

Another cast in place example was the soffit to the elevated floor of the State Tobacco Warehouse, Bologna, 1949. Here Nervi developed an economic moulding form using ferro-cemento with moulds placed on movable metal scaffolding that meant the form could be easily struck and then moved to the next casting location. This did away with the time required to erect and strip conventional timber shuttering, and the finish obtained was so fine that costly stuccoing was unnecessary.

Ferro-cemento is essentially an impregnation and wrapping of closely spaced wire mesh with cement mortar. The cast unit can be 12–40 mm thick depending on the number of layers of wire mesh used.

Gatti Wood Mill, Rome, 1951 is an example of a ribbed suspended floor that used the isostatic ribs formation following the lines of principal stress. Nervi was much influenced by Roman and Gothic architecture and made reference to King's College, Cambridge and the vaulted roof that inspired this design solution.

The same principle can be seen in on page 114 in the Palace of Labour, Turin, which provides a rib layout and change in rib profile, widening and deepening towards each support. The splaying of the reinforcement at the ends of the ribs can also be noted. The mezzanine floor at the same building uses a different rib layout. The smooth radius at beam intersections and slenderness of the rib profiles give the impression the unit is cast in metal not concrete. Such beautiful workmanship could not be surpassed.

Precast work

In 1939 a second competition was announced by the Italian Air Force for six airplane hangars. These massive structures, measuring 100 metres × 40 metres, had openings 50 metres wide by 12 metres high. The construction problem was much the same as before but the doors were now increased in height from 7.9 metres to 11.9 metres. Nervi's company was invited to tender.

Nervi concluded that a reduction in formwork and weight could be achieved by using precast concrete elements joined by welded protruding reinforcement and then pouring concrete to complete the joint. The precast elements were cast in trusses. Nervi considered these structures the most interesting technical and constructional problem he ever worked on. They also exemplified how difficult it was to foresee the true static life of a highly indeterminate concrete structure.

Illustrations show how slender the structures were – and how massive. Sadly, all six were destroyed by the retreating Germans towards the end of the Second World War. The joints were examined and found to be intact, validating the performance of structural prefabrication.

A further circular form of precast structural concrete was the Kurstaal Pavilion: Osita Lido, 1950. Nervi described the performance of the structure as follows:

> The cantilevering perimetric structure is especially interesting from a static point of view. Its stability is given by the reciprocal action between the external tension ring and the one in compression supported by the perimeter columns. In order to resist asymmetric stresses it is essential that this structure be absolutely rigid in plan. This was assured by placing an additional two-way reinforced concrete layer above the precast elements.
>
> The covering of the hall is supported only by the central column. The window mullions function to equalise asymmetric forces due to wind or external loads.

Another example of a circular roof can be found in the Festival Hall, Terme di Chianciano, 1952, an elliptical dome, 32 metres × 26 metres, built with precast elements. The outer two rows have been left open to allow daylight to penetrate into the building.

The Small Sports Palace, Rome, 1955 had a combination of precast and cast in place construction. The edge of the dome required the roof beams to converge towards the support columns but were totally isolated structurally from the inner stadium structures. This provided a clean line and an unhindered view within the building. The beauty of the roof form is evident (see pages 115, 116).

A Brief History of Decorative Concrete

Precast radial roof elements were also used on the Flaminio Stadium, Rome, 1959 built for the Rome Olympics in 1960. They are arranged in a corrugated or alternately stepped form and provide a light interior, with windows in the connecting webs similar to the Turin Exhibition Building.

Ferro-cemento

Nervi described ferro-cemento as follows:

> Ferro-cemento combines the freedom of form of concrete with the strength of steel. It is a 'super' reinforced concrete, ideally suited to the realisation of large structures. It answers two questions: how to manufacture a concrete structure by mass-production methods and how to reduce the cost of forms.

(It should be noted that ferro-cement or ferro-cemento, can be produced by either applying mortar over a mesh of reinforcement or filling moulds containing the reinforcement arrangement.)

For the roof of the Turin Exhibition Hall, Salone Agnelli used a combination of *in situ* concrete beams and ferro-cemento detailing. The parallel roof beams converge into the column supports of the roof, from continuous to discontinuous. The top and bottom beam sections were cast in place. The beam corrugation or wave length is 2.5 metres and the beam sections can be derived from the scale.

In 1949 the Exhibition complex was enlarged, and Salone C was added. After trials a vaulted roof was chosen, with a perimeter slab spanning 10 metres. The steel reinforcement was placed in the lower ferro-cemento ribs and the upper half with reinforcement protruding from the precast units ready to receive the thin slab that will cast in place over them.

Ferro-cemento was also used in the erection of the peripheral floors of Salone C and the main hall roof beams. The beams for the peripheral floors of Salone C were cast on concrete moulds, with the ceiling side in contact with the forms. The concrete forms, in turn, were cast in plaster moulds, which could be shaped exactly. The beams were strengthened by diaphragms at the ends

Decorative and Innovative use of Concrete

and at intermediate points to prevent cross-sectional deformation. The thinness of the sections is a bold choice and remarkable achievement, with the web just 2 cm thick.

The section of beam for the Exhibition Building appears the same as used on Salone Agnelli.

The idea behind ferro-cemento is that concrete can withstand large strains adjacent to reinforcement. The magnitude of the strain depends on the distribution and amount of steel in the component. Nervi pushed this to the limit by surrounding layers of fine steel mesh with a cement mortar. He then tried inserting larger diameter bars, 6 mm and 10 mm, within the centre of 10 or 12 layers of mesh, and clothing this with mortar. The outcome was mortar-coated mesh rather than reinforced concrete; it did not behave like reinforced concrete but rather had the properties of a homogeneous material.

Impact tests were carried out on 28 mm thick 1.5 × 1.5 metre squares, by dropping 250 kg weights from 3 metres. Even if the unit cracked it would be virtually water resistant. This led to the use of the material to construct boats.

The hull of motor sailboat, *Irene*, 1945 was formed by first fixing the layers of mesh to profile and then trowelling mortar to encase the mesh. The boat was a 165 ton vessel of skin thickness 35 mm formed from three layers of 6 mm mesh at 100 mm spacing and then four layers each side of fine mesh at 1.2 kg per square metre. The mortar rendering included 365 kg of pozzolanic cement per cubic metre of sand. The whole weight of the hull was 5% less than the weight of a similar wooden hull and cost 40% less to build. The cost was also quoted as being less than that of a steel hull. After eight years the hull was still watertight.

Nervi made other artefacts in ferro-cemento and then his own boat *Nennele*, a 12 metre ketch, in 1948. The skin thickness of this boat was 12 mm reinforced by seven layers of mesh with a single layer of 6 mm longitudinal bars spaced 5 cm apart.

'Beauty', said Nervi, 'is an unavoidable by-product of the search for satisfactory structural solutions' (perhaps this could also be considered a definition of innovation). His work certainly confirmed his opinion.

3.10.8 Oscar Niemeyer and Brasilia

The history of Brasilia and the architecture of the new capital of Brazil are closely linked to the architect Oscar Niemeyer. It is appropriate to give some background as to how that came about.

The idea of a new capital had been in the making for 150 years. In 1946 administrative work finally started and in 1956 Juscelino Kubitschek, the just elected president, founded Novacap (Nova Capital) to progress the idea of building a new capital in the centre of the country in a depopulated area. Niemeyer was put in charge. Novacap announced a competition for a general construction plan in which all entrants had to be licensed in Brazil. The jury, however, was international and out of 26 entrants they selected the entry from Lúcio Costa.

Thus Costa designed the layout and Niemeyer the buildings.

Oscar Niemeyer Soares Filho was born on 15 December 1907, and became a pioneer in exploring the possibilities for concrete use in construction. He is a staunch defender of utilitarianism, but his buildings are not without a dynamic and sensual form and so do not have the coldness often associated with the material. Many believe the aesthetics of Niemeyer's buildings are such that they are more akin to sculpture than architecture. Others see this same characteristic as one of his faults.

Niemeyer graduated in 1934 from the Escola de Belas Artes as engineer architect and started, without pay, in the studio of Lúcio Costa, a move that later turned out to have been a shrewd and incisive one. In 1936 Lúcio Costa was appointed architect for the Ministry of Education in Rio de Janeiro and in 1939 Niemeyer became leader of the team of architects, which included Le Corbusier as consultant.

Brasilia was founded on a site with no access or service roads. Paths and tracks had to be cut out with bulldozers. There was a huge shortage of skills in the country, and much of the workforce had to be brought down from northern Brazil.

Materials used were almost exclusively Brazilian-made. Reinforced concrete was used for all buildings except the Palace Hotel and the Ministry Buildings. Aggregate had to be sourced from 6 to 37 miles away, and concrete was almost entirely from within Brazil.

Niemeyer set out with a bold objective:

> I favour an almost unlimited liberty of forms, a liberty by no means subservient to the reasons of technique or function but acting, first of all, as a challenge to imagination, to the creation of modern and beautiful forms that may astonish and convey emotions because of their beauty and their embodied creative genius; a liberty that permits, whenever desirable, a mood of ecstasy, of dreams, of poetry.

Illustrations of Brasilia (page 125) show he certainly achieved his objective. Indeed his whole work output testifies to his imagination and the challenge he threw up to the principle of a more functionalistic approach to design.

However, he must have had his critics, as is clear from the following comment:

> Some circles in modern architecture, however, are against such liberty of forms. It is the timid, those who feel safer and more at ease when limited by rules and regulations which leave no room for fantasy, for deviation, for contradiction of the functional principles they adopt and which lead them to accept passively solutions that, repeated again and again, become almost vulgar.

Brasilia had the advantage of setting out on a fresh site with a clear horizon and a thoughtful architect. The building rate was rapid – the city was inaugurated in 1960, from a design start in 1956. Queen Elizabeth visited in 1960 and there is a photograph of her in the foyer of the Itamaraty Palace having just ascended the splendid open plan spiral staircase.

The Planalto, Supreme Court and Alvorada palaces show a restriction of design, concentrating on the forms of the pillars or colonnade. These do not adopt the usual cylindrical or rectangular sections, which would have been far more economical and easier to construct, but have instead curved sail-like proportions (page 124) that impart greater lightness to the buildings, making them appear to almost float or, as Niemeyer wished, 'to touch the ground only slightly'.

[Above] *The Alvorada Palace, Brazil, 1957*
(photo: © Fernanda Hinnig/Dreamstime.com)

[Opposite] *Brasilia Cathedral, Brazil, 1958 (photos: Julian Weyer)*

Decorative and Innovative use of Concrete

A Brief History of Decorative Concrete

The tiny chapel to one side of the Alvorada Palace is ideally suited to concrete construction, being a wrap of a concrete sheet.

The Palace of Congress is an iconic example of an expression of spirit and imagination. It is Brasilia's most powerful and provocative piece of architecture. The building stands between two embankments which were built to provide roof-level access to the palace on the two smaller sides. The open space available has been used with the design to give it an outspoken, monumental aspect by simplifying its elements and adopting pure geometric features. Niemeyer claims that had he followed the critics with their academic minds he would have produced an upright building impeding a panoramic vista instead of this terrace which, with its imposing aspect, provides astonishment to many visitors.

Niemeyer made further comments about the building complex:

> From the viewpoint of architecture, a building like that of the National Congress must be featured by its fundamental elements. In this particular case such elements are the two plenary halls where the vital problems of the country are solved. Our aim was to emphasize their plastic appearance and therefore we transported them onto a huge esplanade where their forms sprout like a symbol of the legislative power.

The Palace of Planalto again has the floating, just touching the ground feeling. It is the seat of the Brazilian government. The name means 'Place of the Plateau'. Brasilia is built on the Central Brazilian Plateau, a central region founded on a plateau, hence the name often used, 'o Planalto'. The structure in front of the building is the Parlatório, built as a place from where the president could speak to the people. It is now only used when a new president takes office.

Brasilia Cathedral is in essence a bundle of hollow parabolic columns formed into a sheaf that represents two hands moving upwards to heaven (page 125). The 16 concrete ribs rise from a concrete ring c.60 metres in diameter; the ring rests on concrete piles.

The Palace of Planalto, 1985 (photo: Julian Weyer)

The floor of the building is 3 metres below the adjacent ground. Entry to the building is via an inclined walkway and the roof is all above ground. Building started in 1958 and the structure was finished in 1960, but completion took until 1970 since funding was via general public donations.

It is said the light blue and green stained windows make the interior feel warm and tropical and perhaps too light for a cathedral. The windows are made from fibreglass in pieces of 10 metres × 30 metres and were painted in 1990 by Marianne Peretti. Bronze statues made by Alfredo Ceschiatti and Dante Croce in 1970 are suspended from the roof.

The Baptistery is oval and has walls covered with tiles painted by Athos Bulcáo. The bell tower in front of the building has four bells, donated by Spain.

After Brasilia, Niemeyer had an active working life, travelling and staying abroad in Europe and then returning to Brazil to lecture and continue his work. He is still working and at the time of writing is approaching 100 years old.

In 1996, aged 89, he created the Niterói Contemporary Art Museum in Niterói, near Rio de Janeiro. The building is sited atop a rock, giving views over the Guanabara Bay. The building is so exotic it is sometimes said to upstage the exhibits. The MAC-Niterói is 16 metres high and its cupola 50 metres diameter, and it has three floors. Around the cylindrical base is a pool that reflects the building as if it were a flower. The structure is in a Modernist style and has a 'UFO' appearance. Access is via a series of sloping ramps in character with the rest of the building.

The Church of St Francis, Pampulha, Belo Horizonte, a series of mathematically determined parabolic concrete arches in the Organic Modern style, was completed in 1943 (page 130, 131). Light enters the building through louvres at the entrance and above the altar. The inside is somewhat subdued whereas the outer walls are overlaid with mosaics applied by artists invited by the architect, including Candido Portinari. The building, like so much of Niemeyer's work, is a combination of his own ideas fused with those of Le Corbusier, Mies van der Rohe and Brazilian Baroque.

*Niterói Contemporary Art Museum, Brazil,
1996 (photo: Rogerio Granato [Rio-Brazil])*

[Above and opposite] *The Church of St Francis, Pampulha, Brazil, 1943 (photos: Geir Haraldseth)*

Decorative and Innovative use of Concrete

Museo Oscar Niemeyer, Curitiba, Paraná,
Brazil, 2002 (photo: Peggy Reimchen)

Decorative and Innovative use of Concrete

Niemeyer's last, but not necessarily final, building
is the Museo Oscar Niemeyer, Curitiba, Paraná, Brazil,
completed in 2002 when he was about 95 years old. The
style is denoted as Expressionist Modern. So much suited
to reinforced concrete, the surfaces are tiled and glazed
with curtain walling. The structure, though rather odd, has
beauty in form and proportions. Like the MAC-Niterói it has
ramp access. The window profile has earned the building
the nickname 'Eye Museum'.

4.1 Colouring techniques

There are a number of ways of introducing colour into concrete. These include selecting sand that has a high fines content, utilising the colouring effect of the fines in the sand. This is practised in the Cotswolds, for example, where oolitic limestone sands provide a characteristic honey yellow colouring. Such products were originally given the name 'reconstituted stone' but are now more commonly referred to as 'cast stone'.

Often either a desired colour cannot be obtained from the fines in the sand or the colouring effect obtained varies. In such cases pigments are needed to ensure a consistent coloured product. Sometimes the uniform colouring appears unnatural to some, but it may be required for certain products. Natural stone commonly has veins or bands of colour variation, and mimicking these in concrete can help to make the appearance more acceptable.

Colouring is also commonly needed when simulating another material such as terracotta or clay bricks, or to provide a vivid appearance. This can be achieved by pigmentation staining of hardened concrete or painting or overlaying the surface with a thin coating of pigment. The latter process is often used on site to camouflage a repair or remedial works.

4.1.1 Pigmentation

Introduction

The prime benefit of incorporating a pigment into concrete or mortar is the consistent colour it generates throughout the material. Pigments can change the image of concrete from a dull, grey material to a vivid and decorative one. The addition of pigments can be particularly beneficial when the surface is subject to heavy wear (industrial floors or roadways, for example) as the surface colour remains consistent. Yellow, blue or red pigmented foam concrete has also been used to encase buried services such as gas and water pipes and electricity cables, so that they are easily recognisable if encountered accidentally.

Definition of pigment and colour

Pigments are fine dry powders, suspensions or slurries insoluble in the application medium (usually water). They are chemically inert in concrete or mortar.

Colour has been defined in terms of three properties: hue, value and chroma.

- **Hue** is a measure of the amount of reflected light.
- **Value** is the lightness or the light-reflecting quality in terms of the lightness or darkness of a particular colour.
- **Chroma** or saturation denotes the richness of hue of a colour and is usually a measure of the purity of a pigment.

Causes of colour variation

It is not only pigment that affects the colour of a concrete or mortar – practically all the other components and dosage levels also have an influence. Pigments work by coating the other ingredients in a mix. Cement is a fine powder and the surface area has to be completely covered if the cement is not to contribute to the overall colour. Thin coatings may allow the grey or white of the cementitious components to influence the appearance of the concrete.

Small variations in pigment additions at low dosage levels can therefore influence the overall appearance, as also can variations in the colour of the sand or a change of cement source.

When a consistent colour is required, for example in precast components or bays in an imprinted concrete driveway, a surplus of pigment has to be added over and above what might be considered the minimum level. This produces a multi-layering of pigment that guarantees a consistent colour.

The colour of both pigmented and unpigmented mixes is affected by the water/cement ratio: the greater the amount of water added, the lighter the colour.

Ways to improve colour stability

Pigments may be colour stable (e.g. iron oxide pigments) or influenced by conditions within or at the surface of concrete (e.g. blue phthalocyanine coloured concrete has a tendency to turn green if acid washed). Carbon black is inert but can be affected by white lime deposits or by leaching out of the surface, leaving a grey appearance.

Colour stability can be achieved only through good concrete practice, including a low water/cement ratio, suitable choice of aggregates and sands (in terms of colour and grading), thorough compaction and sufficient curing. These reduce capillary porosity and increase strength, improving resistance to mechanical abrasion and frost attack. They also tend to reduce the risk of lime bloom, when calcium hydroxide evaporates onto the surface and carbonates to form a white powder.

It is also important to consider the type of pigment used. Carbon black can leach out of concrete, as mentioned above, and organic pigments such as phthalocyanine green have low colour stability in alkaline concrete.

For pigment blends the dominant particle sizes must be compatible. Weathering is likely to affect the finer particles, leading to a change in appearance. It is best to choose aggregates with colouring as close as possible to the intended final colour.

Colour stability has been assessed by both long-term outdoor exposure tests and indoor accelerated weatherometer testing simulating rainfall and sunshine. Other accelerated tests simulate polluted atmospheres. Long-term natural weathering tests give the best guide to overall colour weathering performance.

Types and physical properties of pigments

Pigments have various forms, compositions, particle sizes and shapes. The most common are iron oxides in brown, red and black that have rounded particles of $c.0.1–0.5 \mu m$ and a water absorption of 20–50%. Yellow pigments are based on iron oxide/iron hydroxide and have an elongated profile, a particle size of $0.2–0.8 \mu m$ and a water absorption of 90%.

The table overleaf gives the particle sizes, bulk densities and water absorption properties of a range of commonly used pigments. For comparison, Portland cement is typically

Physical properties of common pigments (from Levitt 1982).

Pigment	Particle size (µm)	Bulk density (kg/m³)	Absorption (ml/100 g)
Red iron oxide	0.1	900	35
Red iron oxide	0.7	1500	20
Yellow iron oxide	0.2 × 0.3	800	50
Yellow iron oxide	0.2 × 0.8	500	90
Black iron oxide	0.3	1100	33
Brown iron oxide	0.3–0.6	900	50
Brown iron oxide	0.1–0.2	1000	30
White titanium oxide	0.2	700	24
Green chromic oxide	0.3	1200	19
Carbon black	0.01	500	100
Green or blue phthalocyanine	0.01	500	N/A (hydrophobic)

$5\ \mu m$ in diameter; the range of pigments can be seen to be c.10–1000 times finer.

Adding pigments to concrete

Because of the variability in bulk density and fine particle size of pigment, volume batching the dry powder can lead to significant variations in the weight being added. Weigh batching is therefore preferred.

Batching pigments as slurries is another way to ensure consistent addition, provided the storage tank is agitated. Incorporating a suspension additive in the storage tank can be beneficial provided the dispersion action does not adversely affect the rheology of the final concrete.

Amounts needed

Additions are normally based on a percentage of the cement or cementitious weight. The typical addition range is 1–10% by weight of cement, and within this the range 3–6% is the most common.

As a guide, an addition of c.1%, which might be used with white cement, provides a tint to the concrete. With an addition of c.5% a definite colouring is produced, and at

10% the colouring is deep and consistent across successive pours or products.

Addition levels for metallic oxide pigments such as red, yellow, black and brown oxides are often in the range 1–10%, whereas those for the finer pigments carbon black and green and blue phthalocyanine tend to be in the range 0.1–1%.

Titanium oxide is a strong whitening pigment that is often used with white cement and aggregates. Addition levels up to 3% are normally sufficient.

Blending pigments

If pigment powder is blended with cement powder, the amount needed to give a specific hue, value or chroma is less than when the pigment is added to the concrete in the mixer (up to 50% less). When relying on a concrete mixer to blend in pigments, a forced-action mixer is practically essential. Pre-coloured cement can be mixed into concrete using a free-fall barrel type mixer but the best option is to blend sand, pigment, cement and part of the gauging water in a high shear mortar mixer before discharging into a forced-action or free-fall mixer to incorporate the aggregate, final water addition and any admixtures.

Green and blue phthalocyanine pigments are hydrophobic and therefore have zero water absorption. To enable wetting and give more control over the colouring effect they have to be pre-blended with about 10 times their weight using silica powder.

Changes in concrete properties from adding pigments

The fresh and hardened properties of concrete can be affected when fine powders such as pigments are added to it. A brief overview is given below; for further information the reader is referred to the current British Standard BS EN 12878:2005.

Fresh concrete properties

Consistence (formally known as workability) is generally reduced by additions of pigments, especially at high loading levels. Inclusion of surface active or wetting agents may be needed with some hydrophobic pigments. This practice is opposed by Levitt (1982, 1998), who suggests it promotes lime bloom. He advocates the use of water-repelling agents to convert the mix to a hydrophobic state, reducing the likelihood of efflorescence, except when using blue or green phthalocyanine pigments which should be mixed with inert filler, including fine silica sand, to invoke hydrophilic properties.

Setting time is apparently not affected when pigments are used at the recommended dosage rates. However, carbon black has been known to produce a flash set.

Air content is not affected by pigments except carbon black and other very fine materials that require a higher dosage of air-entraining agent. Carbon is noted to significantly affect air content when inclusions of pulverised fuel ash (PFA) are combined with air-entrainment.

Bleeding and settlement are generally not influenced by pigments provided recommended dosage levels are followed. However, higher dosage levels of pigment with the use of low C_3A content cements can reduce cohesiveness, which may result in bleed water on the surface.

Hardened concrete properties

Compressive strength is unaffected by normal dosage levels of pigments at the same consistence as unpigmented concrete; if anything, the strength is slightly greater. The exception is yellow iron oxide pigment when included at dosages above 6%, as a result of the higher water demand caused by its fine needle-shaped particles.

An increase in strength has been reported with additions of ultramarine blue pigment, caused by a pozzolanic reaction with free lime.

The current Standard BS EN 12878:2005 allows for a reduction of up to 8% when determined in accordance with the method given in that publication.

Shrinkage is affected as a result of the extra water needed in a pigmented concrete compared with that in a plain mix. Fine pigments such as black and brown iron oxides again have the most significant effects increasing water demand and reducing air entrainment.

Some pigments in the USA are blended with air-entraining agents. Improvement in dispersion is achieved with the aid of a superplasticiser that will also reduce water demand and thereby shrinkage.

Durability of pigmented concrete

Freeze–thaw resistance is a function of water content and pore structure. Carbon black has been found by some to have a lower or poor freeze–thaw resistance. Others have compared freeze–thaw resistance of concrete blocks and found that if the block strength is above 60 N/mm^2 the resistance was satisfactory. This is no different from unpigmented concrete.

Permeability of pigmented concrete is related to water demand as in unpigmented concrete. If pigments in suspension are used with water repellents, this may reduce the permeability in proportion to the water/cement ratio.

Further guidance

The following guidelines on the use of pigments are offered by Levitt (1982).

1 Formwork or moulds should be of good quality and finish. High gloss paints and high gloss plastic mould surfaces should be avoided as they promote hydration staining. Ideally, all finishes should be matt.

2 Mould release agents should be water-in-oil emulsion cream or chemical release agents. Ordinary mineral mould oils and oil-in-water emulsions promote streaking and staining.

3 Pan-type mixers are preferred to tilting drum but whatever type of mixer is used it should be well maintained with blades properly set, and thoroughly cleaned at the end of working periods or when there is a change in the mix requirements.

4 Pigment should always be weigh batched or, when in the form of slurries or suspensions, weigh or volume batched by an approved dispenser. Accurate control of concentration is always important.

5 Dry pigments or blended cement should always be added with the nominally dry ingredients first and mixed for 1–2 minutes. Suspensions and slurries should be added with the water after the dry mixing period.

6 All pigmented concrete should be weigh batched; it is advantageous in the case of volume batching that bulk densities and moisture contents are known so that proportions can be corrected from mix to mix.

7 Compaction should be as effective as possible.

8 Finishing tools should be wooden or acrylic or, in the case of steel, stainless to prevent staining.

9 Curing should be consistent to maintain a uniform colour and pigmented materials should not be subjected to extreme conditions during the first few days. Membrane coatings can be advantageous if the environment cannot be controlled but polythene or similar types of covers are not recommended as they cause condensation and may stain the concrete.

10 Coloured concrete and mortar can be finished in the same way as ordinary concretes, but note that blue phthalocyanine pigmented cast products may turn permanently green if etched with hydrochloric acid.

Final remarks

Lime bloom is a significant problem affecting the appearance of pigmented concrete. The risk of its occurrence can be reduced by keeping water content to a minimum and incorporating a stearic acid water repellent at c.1% addition by weight of cement. Semi-dry cast stone should incorporate 1–2% by weight of cement.

Carbon black can invoke a less workable mix but the water content should be kept to a minimum.

Normal Health and Safety precautions should be employed when working with pigments, such as wearing protective clothing and employing air extraction equipment to keep dust to a minimum. Handwashing using baby soap is recommended as the fine grain penetrates into the contours of the skin.

Where a pale colour is required it is better to use a large amount of a weak pigment, or dispersion of a pigment, than a small amount of a strong pigment in order to obtain a consistent appearance.

4.1.2 Staining and dyes

Introduction

A variety of materials have been used for many years to stain concrete. In the 1930s wood stains were applied on dried external concrete that had been pre-treated with a solution of fluorosilicate of zinc and magnesium to neutralise the alkali. Several applications were often needed to get a satisfactory depth of colour. Internal work was then oil-wax coated to bring out the colours. Some wood stains will not withstand sunlight and should not be considered for external work.

Perhaps surprisingly, aniline dyes have been used in a solution of wax in petrol and applied and finished as the wood dyes above. These were only used internally.

Another cheap method has been to brush the surface with a solution of ferrous sulphate. This is a green-coloured powder or crystals, and the resulting buff shade is from a chemical reaction between the sulphate and the cement. Stronger solutions give deeper shades, as do repeat applications of one concentration.

Solution strength varies between 10 and 25%. A 10% solution can be batched by adding 0.5 kg of ferrous sulphate to 5 litres of water; this will treat c.25 m² as a single coat. Even application is necessary, and care should be taken not to overcoat areas already treated, to avoid further darkening of the surface.

Four or five days drying time should be allowed between coats, and the final appearance judged three or four days after applying the last coat. The older the concrete the less the colouring effect obtained. Copper sulphate can also be used to obtain shades of green.

Types of staining product
There are two types of staining product:

- acid-based reactive stains
- solvent- or water-based dyes and tints.

Acid-based reactive stains
Staining is now used on new concrete as well as a means of renovating old concrete. Reactive types of stain are metallic salts in a water–acid solution. These acid-based stains react with calcium hydroxide in the concrete surface.

Solvent- or water-based dyes and tints
Dyes and tints do not react with concrete and are not applied with an acidic solution. Dyes offer colours that are not available in reactive stain form and can combine with stains to produce a more uniform colouring. They can also be used to lighten concrete, which is not possible with reactive stains.

Dyes may be organic or inorganic and can therefore be subject to ultraviolet bleaching.

Stains should not be confused with paints. Unlike paints they become permanently infused into the surface of the concrete as the acid solution dissolves the calcium hydroxide and limestone aggregates to open the surface of the concrete.

Concrete usually has to be at least 14 days old before it can be stained. However, for blue, green and gold colours new concrete needs to be 30–60 days old and well cured before a satisfactory result can be achieved.

Application is normally by spray followed by brushing into the surface. This helps the acid dissolve and react with the surface by displacing eroded surface calcium particles, allowing new areas to take up the solution. Effervescence is always noted when acid acts on calcium compounds.

After about four hours the surface is then washed with a detergent to remove surplus stain and residue. If further staining is needed, or other colours are to be applied, the surface has to be dried first. Only when washed can a surface be assessed to ensure the desired effect has been achieved.

Techniques in staining and dyeing concrete
There have been major developments in staining techniques over the years, especially in the USA. It is a detailed and intricate the process, with numerous different and sometimes spectacular effects possible in the hands of skilled and trained operatives.

The finished surface is always a function of the skill and knowledge of the applicator and is influenced by variations in absorbency. Variation in stain take-up can be used to provide an interesting and aesthetically pleasing appearance.

Dyes are not active and should preferably be used after staining to enhance stains. They can be mixed to provide a precise colour and applied with brush or spray or paper, cloth, etc.

It is imperative to seal the finished floor with an acrylic sealer. The amount applied will determine the amount of sheen produced. Fine mist spraying of successive thin, hazy coats allows each coat to dry quickly and the amount of gloss to be controlled.

The following special effects can be achieved:

- Designs can be achieved by sandblasting after staining through a stencil.

- Applying stain with torn paper or rags generates veins or cracked edges.
- Fertiliser pellets dissolve and provide a darkening effect in patches.
- Eyedroppers allow stain to be dropped in local areas.
- Applying stain over aluminium swarf intensifies the pigmentation local to the area. Swarf is removed after the reaction has occurred.
- Cat litter can be used to absorb fresh stain, producing a local lightening effect.
- Wax and acrylic sealers can be use to mask areas.
- Stain can be sprayed over ferns or other leaves, or they can be dipped in the stain and then laid on the surface.
- Two runs of stain can be mixed on the concrete by pouring from containers.
- Brushing the surface or wiping with white paper pads can produce swirl patterns.
- Working lines can be drawn using permanent marker pens and stain applied up to and just over the mark.
- Stain can be sprayed through lace curtains or open weave cloth.
- Stain dropped into cut lines and then blown out with an air-line can produce a fern pattern.

Other considerations

Preparation of the surface is most important. Sample areas should be large enough to convey the appearance intended, and preferably located alongside the job on spare areas cast for that purpose so as to utilise the same concrete and finish.

It is often not appreciated that the final result, in terms of variations in the intensity of the stained colour, will depend on the precise permeability and absorption of the concrete. Trials on samples of the actual concrete to be stained are therefore mandatory.

4.1.3 Precast pigmented concrete

Precast pigmented concrete is essentially covered in section 4.1.1. However, it is appropriate to mention the kind of finish that can be achieved when vivid pigmentation is chosen.

Colour varies in intensity over each mosaic piece, due to various factors. Uniform, intense colour is very difficult if not impossible to achieve with pigmented concrete.

4.1.4 Dry shake toppings

Introduction

Dry shake topping finishes are blended, dry bag materials, consisting of cementitious binders, sands, admixtures and pigments. They are available in a limited standard range of colours. No Standard covers the product.

They are also used to:

- improve the abrasion resistance of the concrete surface
- provide a pleasing appearance to floors
- cover any surface fibres when steel fibre reinforcement has been used.

A range of dry shake aggregate types has evolved, falling into three categories:

- metallic and metallic alloy sands
- natural mineral aggregates
- synthetic minerals which are non-metallic.

The metallic aggregates are most hard wearing, with ferro-silicon and aluminium oxide being almost as hard as diamond. Bulk densities vary considerably over the range of the three aggregate types and since there is no Standard for these products, grading envelopes have been developed by each manufacturer. An inappropriate product grading, combined with the cement addition, can cause delamination in service.

Considerations during application

Dry shakes are spread over freshly placed and screeded concrete floors. They can be applied by a telescopic spreader or by barrow spreader or shovel. The objective is to apply sufficient powder uniformly over the surface. The powder is then power trowelled into the wet concrete.

Cement in the dry shake is hydrated by bleed water migrating to the fresh concrete surface. The mix therefore has to provide sufficient bleed water to completely hydrate the applied dry shake. Too much or too little and the surface may fail in service.

Curing

Curing starts immediately after the final trowelling and has to be continuous for seven days (or longer in cold weather). This is best achieved by a spray-on membrane or by overlaying a waterproof sheet.

Defects and blemishes

Defects can occur and it is wise to have agreed methods of repair. Any grinding of high spots after the finish has been completed may expose plain underlying concrete.

Delamination occurs when areas fail to bond. The thickness is normally 3–5 mm and delamination can be detected by tapping. Any reinstatement method will need to be tested first to ensure it is acceptable.

Cracking is not uncommon on large jointless floors that employ dry shake finishes. Crack filling with low-modulus epoxy resin is the most effective solution.

Crazing is produced when excessive trowelling brings to the surface a layer of laitance that dries more quickly than the bulk of the concrete. Restraint from the underlying concrete leads to differential drying shrinkage and surface crazing. Crazing is shallow and not normally detrimental, and is found quite commonly on this type of finish.

Pinholes and pop-outs are cause by small aggregate particles or dry agglomerates of dry shake working loose. It is better to leave them than try to carry out repairs, as the latter can be more noticeable.

Larger holes (5–10 mm diameter) and impact damage are best repaired by cutting out and filling with a high-strength mortar or epoxy resin.

Dry shake finishes are often found in large DIY superstores.

4.2 Texturing techniques

4.2.1 Concrete requirements

Before a satisfactory finish can be produced from any concrete, an appropriate formulation is required to enable the surface to be moulded or formed. Much emphasis is placed in specifications on the strength, durability, fire performance and even thermal characteristics of a concrete, but little attention is given to ensuring it has the capability of being finished.

There are six characteristics of a concrete mix that affect ease of finishing:

Recommendations for concrete that should produce a good finish.

1	Cement content	Minimum 350 kg/mm^2
2	Sand content	Not more than twice the cement content
3	Total aggregate	Not more than six times the cement content
4	Coarse aggregate	20 mm aggregate, not more than 20% less than 10 mm
5	Consistence	Target of 50 mm slump
6	Water/cement ratio	Target of 0.5 or less

It is worth noting that although the target slump is 50 mm a ready mix supply could have a slump value between 30 and 110 mm and comply with that value. When finish is important, a special agreement should be reached between the concrete supplier and the purchaser to limit the consistence level by whatever means, such as adding admixtures on site to a lower target slump concrete, rather than making do with a variable and possibly too wet 'normal' concrete supply.

The above recommendations have been found to produce concrete with minimal surface blowholes and little bleed water migration towards the form face. Where formwork release agent is required, consideration should be given to the following table and an appropriate type chosen (e.g. type 5).

Other considerations are the rate of pouring in terms of height rise up the shuttering. A rate of at least 2 metres per hour is appropriate to prevent colour changes between lifts that are ideally kept to 0.5 metres high.

The top of a lift should be left to accumulate any bleed water and for plastic settlement to occur. Re-vibration or working the surface before set takes place will ensure the face is compacted and heal any settlement cracks. The top of the lift should then be wrapped in plastic, ideally with damp hessian, to prevent the top of the pour drying and darkening the surface.

Deep lifts cast against absorbent shutters will vary in colour if the shutter is not sealed before use. All types of shutters clearly need to be grout tight, even if the face is to be subsequently worked or aggregate exposed later. Sand grading is critical to avoid segregation and bleed

Formwork release agents categorised into types and appropriate uses.

	Type	Notes	Application
1	Neat oils (no additives included)	Not for visual concrete. Can promote air entrapment	
2	Oils with surfactants	Reduced air entrapment. General purpose use including on steel shutters. Affected by heavy rain	Soft brush or spray applied. Typically 20–30 m² per litre
3	Mould cream emulsions (water in oil with a surfactant)	Reduced air entrapment. For all uses except steel, and suitable on absorbent formwork	Disperse well before use. Brush or sponge applied. Avoid frosty weather
4	Water-soluble emulsions (oil in water)	Not for visual concrete. Causes retardation. Promotes dusty surface	
5	Chemical release agents (chemical compounds in a thin oil)	Reduced risk of surface blowholes. Good with impermeable formwork. Let dry before use. Rain resistant. Excess retards	Spray or sponge applied. Typically 35–50 m² per litre. Just a hazy spray application
6	Barrier paints and varnishes (impermeable coatings)	For surface preparation. Release agent needed	Follow manufacturer's instructions
7	Waxes (in solvents)	For surface preparation of concrete and other surfaces	Follow manufacturer's instructions

water migration. A little air-entrainment can be used when sand grading shows a lack of fines and alternative fines are not available. Too high a cement content can generate damaging thermal gradients that can induce differential cracking, so increasing the fines content by adding more cement should not be considered an option.

4.2.2 Form finishes

Formwork provides the shape and texture to encased concrete. The examples below illustrate how varied are the forms and textures that can be achieved; the possibilities are limited only by the imagination of the designer.

The Chen Residence in Kew, Victoria, Australia by Ivan Rijavec is an example from 1996 that recalls the 1960s and 1970s white building style. It bursts with radial geometry. Rijavec describes it as

> a box compressing unruly curves which squish through the container to burst out at each end, and finding expression on the front and rear façades.

The surface is plain and smooth but the form is made intriguing by the shape and profile of the curved entrance tower.

Canary Wharf Station, part of the Docklands Light Railway, has appeared in many books and magazine articles as a superb example of the ability of concrete to provide curvilinear form that does not require texture. I find it a little amusing when this station is cited as a fine example of the possibilities of concrete as a facing material, as I was employed to investigate the dissatisfaction expressed at the time over the mottled and non-uniform colour of the surfaces. This photograph, as well as the other examples, shows the slight variations in the finished colour and lighting effects that give the surface of concrete an appealing individual appearance.

Acceptance of a less than perfect concrete surface as a result of modest resources and difficulties in quality control led Louis Kahn to 'ennoble' rough concrete by demarcating the joint lines between concrete pours with marble inserts, as in the Assembly Building in Dacca, Bangladesh.

Canary Wharf Station, London (photo: David Groom)

A similar juxtaposition of angles and lines is found in the 1980 and 1994 remodelling of the original 1963 Sheats/ Goldstein House by John Lautner. This shows concrete furniture and fittings with irregular angles and leather cushions, but with the concrete smooth as intended.

Tadao Ando's Koshino House (http://en.wikiarquitectura. com/index.php?title=Koshino_House) has a regular pattern of formwork panels and through tie holes, but with curved surface profiles that are dematerialised by the raking sunlight.

[Above] *Sheats Goldstein Residence, Beverly Hills, California, living room chair (photo: John Lautner and Duncan Nicholson)*

[Right and Opposite] *Sheats Goldstein Residence, Beverly Hills, California, Outdoor furniture (photos: Duncan Nicholson)*

[Below] *Sheats Goldstein Residence, Beverly Hills, California, Skyspace roofdeck bench (photo: Duncan Nicholson)*

Decorative Processes and Techniques

Cast ferro-cement panels, using the Ferro-Monk system were employed in the ceiling and protruding first floor soffit of the Schlumberger Research Laboratory, Cambridge, and recall the work of Nervi in the 1940s and 1950s employing 'isostatic ribs': smooth clean lines, much suited to and ideally formed against a GRP (glassfibre reinforced plastic) mould face in cast quality concrete, together with a uniform finish characteristic of overcoating.

It is said that Ove Arup 'found greatest pleasure in simple structural concepts that were elegantly expressed'. The Kingsgate Footbridge, Durham, completed in 1963, was one of these 'structural concepts'. The Concrete Society voted this bridge the winner in its Annual Society Awards, giving it a Certificate of Outstanding Performance in the category of Mature Structures. The timeless quality

Decorative and Innovative use of Concrete

[Left, below, opposite and following page] *The Kingsgate Footbridge, Durham* (photos: Phil Henson, http://www.flickr.com/photos/24583241@N04)

Decorative Processes and Techniques

and considerate detailing have contributed to sympathetic weathering, and the original colour is now more mellow, with some lichen growth and water staining introducing variation over the surface. Maintenance-free except for cleaning out the waterspouts, concrete was chosen after discussions with the client and the Royal Fine Arts Commission, because it enabled the most sympathetic details and textures to be produced for the complex shapes and intersections.

There is an interesting contrast between the opinion of Ove Arup that the structure was essentially a simple concept with that of the client who saw it as complex shapes and intersections. No doubt much of the success of the project can be attributed to Arup's spending many hours at the site perfecting every detail of the design.

Less common off the casting lines today are finishes formed against rope and rough sawn timber. The striated, textural finish is produced by parallel lengths of manilla rope, usually 20 mm diameter and spaced at 50 mm centres. A tapered batten is used top and bottom to produce a plain edge and the rope passes through holes that act as spacers and a fixing. Panel pins are hammered through the rope at 250 mm centres to fix to the timber back shutter and the lengths are looped at each end to assist removal. The back shutter is coated with release agent before the rope is fixed and the rope soaked with water before the concrete is cast.

After 48 hours the back shutter is pulled away, leaving the rope in the concrete with the panel pins sticking out. The loops in the rope allow a winch to pull each length free of the concrete. Up to 12 uses may be had from each set of ropes.

Boardmarked concrete was much the fashion in the 1950s and 1960s. Many examples have failed the test of time, but in the right location it has proved successful. Deal, pine or Douglas fir are used. Wood with an appropriate weathered raised grain is selected, or the surface is roughened by sandblasting. A stepped finish is produced by using strips of differing thickness with the protruding edge cut to a chamfer to allow the form to strike without ripping away the concrete arris.

One benefit of using a boardmarked finish is that minor defects or blowholes may not be so noticeable. Boards must be carefully selected, with any variations in the texture of the boards spread uniformly over the surface. Fixing of boards is through the back of the support sheet. The facing is then given a conditioning soaking to cause the strips to swell and form a grout-tight seal. This assembly is then wrapped in polythene until used. If a strip has to be replaced, it is likely to produce a noticeable difference in the concrete unless it is given a thorough soaking before the panel is used again.

An interesting personal account on the 'virtues' of boardmarked concrete was given by George Perkin in *Concrete*, June 1990, Vol. 24, No. 6. He states:

> It was about the unfortunate vogue for exposed in-situ boardmarked concrete that swept the country in the fifties and sixties. Alas, this regrettable technique was used widespread in public places where it was most seen – in underpasses, overpasses, walkways, stairways and entrances to tower blocks, flats not to mention the fashionable buildings of the day such as the Hayward Gallery and the National Theatre on London's South Bank.

His caption to an illustration of a corner of the National Theatre reads:

> The technique that tarnished the image of concrete: boardmarked insitu concrete used for the external walls of the National Theatre.

Perkins comments further:

> … talking of the South Bank, the greatest and most disgraceful eyesore to be seen today in London is 'cardboard city' beneath the in situ concrete terraces of the South Bank, where whole colonies of vagrants are now camped in what amount to dormitories. As an example of how 'image' can work, it is possible to look at this appalling scene with the two unhappy circumstances of awful-looking concrete and vagrancy and blame the latter on the former.

I will leave the reader to reflect on the benefits and possible drawbacks of the finish, but soiling associated with boardmarked concrete is largely due to the adsorbent surface taking in airborne pollution with surface rainwater. This can be prevented to some degree by treating the surface with silane or siloxane or even an acrylic paint to shed rainwater and wash off the pollutants. Those of us who attended the Cement and Concrete Association Training Centre in Slough will remember boardmarked finish used internally.

Foamed plastics such as polystyrene and rigid foamed polyurethane have been used to produce one-off designer sculptural forms for concrete. An Illustration (see page 212) shows a profiled pattern obtained using a profiled concrete mould face lining. The profile in this case is relatively slight and can be seen as similar to the GRP formwork liner used to cast the large abutment wall (see page 215). The latter should be considered for repeat casts whereas polystyrene is ideal for one or more varying designs of special pattern profiles, or if lettering is to be incorporated.

Plywood can also be used to produce a decorative form. Examples specially designed and prepared by William Mitchell, the most notable exponent of designer concrete surfaces are shown in chapter 5.

A real benefit in using polystyrene or any foamed plastic is the deep profile that can be relatively easily achieved. Liesbeth Berkers has sculptured expanded polystyrene into formwork for panels for over 20 years. William Mitchell has used polystyrene form liners to create an infinite variety of shapes and textures, including geometrical shapes and deep undercut relief patterns of puzzling complexity. Unlike other rigid form liners, demoulding is not a problem since the polystyrene can be broken away from the concrete and discarded (nowadays dissolved with a solvent and recycled).

Due to the difference in density, polystyrene mould facings can be displaced by fresh concrete so they require secure fixing to the backing. Form, profile and shaping can be produced with a knife, saw or hot wire. Any undercut has to be considered since it may trap air that will leave a void on demoulding. Nowadays concrete is best made highly workable with a superplasticiser to assist compaction rather than using vibrators that may damage the liner.

A type of high relief formed concrete, which has weathered better than boardmarked finish, is produced by casting a ribbed profile and then hammering away the

[Below] *William Mitchell; Hatfield Water Company (photo: William Mitchell Designs)*

[Opposite] *Detail of a knapped finish. Salters Hall, London (photo: Phil Henson, http://www.flickr.com/photos/24583241@N04)*

edges alternately up the run of the rib. An example is Salters Hall, Fore Street, London EC2, awarded a Certificate of Excellence in the 2004 Concrete Society Awards. The close-up view shows the way the ribs are broken to form a wriggly profile and not just to expose the aggregate.

This finish was first used at the Elephant and Rhino building in London Zoo, 1962–65 by the architect Sir Hugh Casson, and the concept is still considered appropriate today. The West Retaining Wall of the Pen-y-Clip Tunnel on the A55 North Wales Coast Road is another fine example of what is now known as hammered ribbed finish. This 1993 application has apparently generated more aggregate exposure than the 2004 Salters Hall example due to the type of aggregate used.

It is prudent to cast test panels alongside a structure so the hammering technique can be demonstrated and the appropriate hardness of the concrete determined before the operator starts work on the structure proper.

4.2.3 Imprinted concrete

The technique of imprinted concrete emanated from California about 50 years ago and is now used widely in the USA. Acceptance has been slower in the UK but take-up has now grown into a significant share of the domestic drive and patio market. Notable commercial applications are holiday camps and seaside promenades, where the colour and textural possibilities of the technique are exploited to the full.

The durable nature of the finish can be judged by the fact it is used for road surfacing. In the hands of expert installers the finish can be quite dramatic, with simulations of various stone textures and colours possible as well as brick and even wooden decking.

[Above and left] *Cobble finishes (Decorative concrete products manufactured by Roy Hatfield Ltd)*

[Opposite] *Sealing the cobbles (Decorative concrete products manufactured by Roy Hatfield Ltd)*

Decorative and Innovative use of Concrete

Decorative Processes and Techniques

First attempts used a more crude method in which a pre-finished concrete slab was cut into rectangular brick-like units by a tool known as a 'cookie cutter'.

The imprinting process

The examples of a typical drive show how different finishes are achieved.

Sub-grade is excavated down to a level c.200–220 mm below the intended finished surface level. It is important any future services are considered, with ducting installed as necessary to allow electrical cables or other services to be run under the concrete. Even existing services such as water mains can be encased in ducts, to both protect them and allow future replacement without disrupting the finished arrangement.

Consideration must also be given at this stage to surface water drainage. Falls and levels must be planned, and drainage pipework, surface gullies and channels located to remove surface water. Sub-base thickness must be in accordance with the ground conditions but ideally should be not less than 100 mm for a surface to take private cars.

Once the sub-base has been compacted and set to levels and falls, it is prudent to cover it with a layer of sand (c.30–50 mm) to form a flat surface over which a polythene slip membrane can be laid. However, if the sub-base is compacted to form a relatively closed surface then the sand layer may be unnecessary.

Slip membranes have the advantage of retaining water within the concrete in summer when the sub-base is dry and likely to absorb water from the concrete, and doing the reverse – preventing entry of water into the concrete – in winter when the sub-base can be very wet. However, the prime function, together with the sand-blinding layer, is to prevent restraint of the concrete slab and allow slippage as the concrete sets and shrinks. Concrete has a low resistance to tensile forces as it sets and hardens, and any restraint from the sub-base can initiate cracking that will open further as the concrete matures.

It is usual to lay c.75 mm of concrete for a patio and 100 mm for a domestic driveway (or more if heavy vehicles are likely to travel over the area). Placing and compaction are carried out quickly, leaving sufficient time for colouring and the imprinting sequence.

Decorative and Innovative use of Concrete

A range of proprietary stencil designs (picture: courtesy of Creative Impressions)

FLAGSTONE

OLD CHICAGO BRICK

RUSTIC BRICK

RUNNING BOND

ASHLAR SLATE

FACE BRICK

STAR COMPASS

JUMBO BRICK

BASKET WEAVE

COBBLE BORDER

SQUARE TILE

SOLDIER COURSE

PINWHEEL

OLD ENGLISH COBBLESTONE

DIAMOND TILE

COBBLE CIRCLE
Dia: 162cm

FISH SCALE

LARGE CIRCLE
Dia: 205cm

The stencil is carefully removed leaving a masked outline of the joint design (photos: courtesy of Creative Impressions)

Decorative and Innovative use of Concrete

Example of a typical imprinted concrete driveway.
Perhaps too shinny and over colourful for some!

Decorative Processes and Techniques

Imprint and texture can be applied to vertical, sloping surfaces as well as horizontal. (photo: courtesy of Creative Impressions)

Decorative and Innovative use of Concrete

Once a level slab has been formed, the procedure starts with the application of the base colour, known as colour surface hardener. This is dusted over the surface and trowelled to an even colour, denoting wetting of the powder by surface water. The second powder, known as the release agent, is then applied in the same way. Release agent is a combination of talc, pigment and additives. The talc allows the printing mats to be removed without them adhering by suction, preventing disruption of the concrete surface, while the inclusion of pigment is a secondary antique effect. Sometimes a liquid, non-pigmented release agent is used.

When the concrete has attained a suitable surface stiffness it can be imprinted. This is determined in various ways, including pushing a finger into the concrete to see if the surface will retain the probed hole without infilling. The time to this stage depends on the type and class of concrete used and the weather conditions.

Imprinting can be carried out in several ways. The first method used a giant 'cookie cutter' to imprint a joint pattern into a flat surface. This would round the edges of the cut stone profiles, but often polythene sheeting was laid over the concrete and the imprinting done into or through this to round the edges even more.

The rounded shape achieved by imprinting with polythene means this process is ideal for simulating English street cobbles. The polythene can be left until the concrete is hard and acts as a curing membrane, avoiding the need for a release agent and protecting the surface from rainfall.

The most common process currently used is to imprint with flexible, synthetic rubber mats that have been cast against a master surface. They fit together and form a surface that is impacted into the concrete using a mat impacter. Sometimes they are trodden into the concrete.

When a row has been installed the first mat laid is taken up and a second row laid abutting the imprinted pattern in the first row.

For applications where a good deal of repeat work is planned, such as footpaths, a roller imprinter is quicker.

Individual imprints such as smooth perimeter patterns or special areas can be extruded or cut into the surface.

Shrinkage or movement joints
After concrete has set, hydration of the cement initiates shrinkage. Concrete is liable to greatest shrinkage at earlier ages, when it is also least able to accommodate restraint. Any shrinkage stresses that are generated at earlier ages cannot usually be accommodated by the low strain capacity of the concrete, and cracking can occur.

To accommodate shrinkage and control cracking, movement joints are cut at restraint points such as the corners of buildings, from the corners of drain hatches, at changes in width and across long runs of paths. The aim is generally to split the whole area into smaller areas (c. 20 m^2) with a length to width ratio not exceeding 1.5. Cutting is carried out either whilst the concrete is still plastic but not fluid or soon after it has set, ideally within 48 hours of placing.

When plastic, concrete can be cut with special hand cutting trowels called 'groovers'. When hard it has to be sawn. The older the concrete the deeper the cut has to be to prevent adjacent crack propagation. Once the surface has been cut the reduced depth under the cut ensures any subsequent shrinkage propagates cracks through the remaining depth under the cut, leaving the surface crack free. At least, that is the objective!

Washing down and finishing

After a few days the surface will be strong enough for surplus release agent to be washed away. It is customary to leave just enough to fill the grain and jointing parts of the imprint to provide an antiquing effect. Cut joints are then filled with sealant, and the whole surface, when sufficiently dry, is coated with a fine spray of acrylic sealer.

Imprinting concrete mix

Imprinted concrete is normally air-entrained and includes polypropylene monofilament fibre, which improves resistance to plastic shrinkage cracking. The level of fibre inclusion is rather low (c.0.9 kg/m^3).

As concrete stiffens any fibre inclusion will disperse stresses that accumulate as the concrete sets and shrinks, giving rise to many close-spaced fine cracks that are normally unnoticed, instead of one single wide crack. Once concrete has set polypropylene fibre has little control over shrinkage strain due to the difference in modulus between the concrete and the fibre. Initiated cracks may open further unless that movement is accommodated by joints.

Concrete content should conform at least to PAV1 specification and should ideally have a cement content of not less than 320 kg/m^3, 10 mm aggregate and a water/cement ratio not greater than 0.55. PAV1 is air-entrained. This conforms to a designated mix suitable for driveways as denoted in BS 8500-2:2002: *Concrete – Specification for constituent materials and concrete*.

Problems that can occur with imprinted concrete

As with any concreting process, it is important to know the material before any work can be carried out with confidence of success. If concrete is not stiff enough, imprinting will create hollows in the surface. If the concrete is too stiff, other defects may be imprinted, requiring the job to be aborted, the concrete to be removed before it has hardened (or certainly the next day) and the area to be started again.

Decorative and Innovative use of Concrete

The following are some problems that can occur through lack of care or knowledge on the part of the installer:

- Insufficient imprint depth. Mats not impacted fully into the concrete.
- Imprinting after the surface has started to stiffen whilst the bulk of the concrete is still plastic. The imprinted profile shows edge cracks.
- Surface crazing occurs when colour surface hardener reacts with free water near the surface and desiccates the top layer, especially when sunshine and/or wind plays over the surface. Free water from within the slab is not available to re-hydrate the surface and crazing appears.
- If a sealer coat is applied before the surface has dried sufficiently, lime bloom may occur under the sealer and the sealer itself may re-emulsify.
- Local areas of replacement are often very obvious due to the contrast in texture and colour.
- Any feature formed must be a correct shape.
- When movement joints are cut across a feature panel it is prudent to ensure the feature and the cut joint are aligned, otherwise the result is disappointing.
- If too much release agent is left in place, charcoal, it will peel off with the sealer exposing the base colour surface hardener. The whole area then has to have the sealer and excess release agent removed and redone.
- The top illustration on page 170 shows a surface where an excessive amount of sealer has been applied, leading to sealer, including lumps of black pigment, lying in pools in the joints. The surface had blisters of entrapped air that 'popped' under foot traffic.
- Inspection hatches are best imprinted out of the frame and the lid placed in the frame the next day.

Obviously the above are not the norm since most installations are carried out correctly. It is worth ensuring, however, that any installation is entrusted to a workforce who can demonstrate ability to the level required for a particular job.

[Above] *A surface where an excessive amount of sealer has been applied (photo: Author's collection)*

[Right and below] *Even wood can be replicated using imprinted concrete (Decorative concrete products manufactured by Roy Hatfield Ltd)*

Decorative and Innovative use of Concrete

Convincing effects can be achieved when the workforce know what they are doing. The cobbled dockside in King's Lynn was imprinted for a film set. Realistic wooden surfaces can be produced and with a little thought and planning an interesting arrangement can be produced from a slab and gravel pathway by laying concrete against side shutters, texturing the surface as one, then cutting out the joints and filling with a contrasting mortar.

Simple domestic imprinted drives that are appearing all over the country can with a little thought be made more individual and interesting. However some surfaces are easier to achieve than others, for example, a slate ashlar surface is not for the beginner because the imprint depth is shallow and any wavy surface would easily be noticed in the reflection.

4.2.4 Stencil pattern concrete

Stencil finishing can be carried out over both old and new concrete. Old concrete surfaces can be transformed to serve a new use, or repaired surface damage can be camouflaged. Brick, block or stone simulations can be stencilled onto surfaces and installed on vertical and elevated surfaces alike, as well as on horizontal floor slabs.

New concrete will require installation of movement joints to accommodate long-term shrinkage and thermal and moisture movement. Old slabs may still benefit from installation of movement joints; although shrinkage is less likely, thermal and moisture movement may be significant on thin slabs.

The stencilling procedure

1   Old surfaces need to be cleaned with acid to remove dirt, oil and lime bloom, to assist bonding of the new surface to the old. New surfaces need to be trowelled to the correct profile, levelled, and joints brought through to the finished surface.

2   All adjacent surfaces must be covered in polythene or sprayed with a protective gel to reduce cleaning time and prevent staining from overspray.

3   Bonding base coat is then either spread or sprayed over the surface. This is usually a neutral colour.

4   Top coat is mixed by adding the powder to the activator in a mixing bucket and the two blended using a paddle mixer.

5   Stencil sheeting is laid over the surface and held in place with tacky putty such as Blu-Tack®. Tailoring a stencil sheet or using special border sheets provides a more solid edge.

6   Spray-applied top coat tends to produce a textured finish but a smoother, 'knockdown' or semi-smooth surface can be obtained by subsequently steel trowelling the surface.

7   When the surface is firm and dry the stencils are removed. A small sample section in the pattern can be used to judge the correct time for removal. Surfaces must not be walked on until the small fractured chips which will have fallen from the stencil have been removed with a mechanical blower.

8   Surface sealing is as with conventional decorative concrete, using spray-on sealers.

Most stencil finishes are relatively smooth. Mortar joints are formed by the base coat and that is why it is usually a neutral colour.

4.2.5 Sandblast finishes

Exposed concrete finishes, as discussed further in section 4.2.7, involve removal of surface laitance. This section considers surface removal but in a different way: its use on contrasting surfaces. Here the possibilities of concrete are exploited to the full to form a profile that would not be possible with other materials at the same cost.

A smooth concrete surface from a GRP mould requires little abrasion via sandblasting to remove mould lines, round off corners, introduce more texture and roughen a surface to complement the casting.

In a contrasting example (see page 9), sandblasting has transformed a flat, smooth surface on a government building in Oslo by drawing a thin line tracing the outline of a Picasso design.

4.2.6 Polishing

Polishing both smoothes the surface and removes as much of the surface layer as required. On floors, polishing is normally carried out using first coarse diamond grit embedded into planetary cutting pads on the grinder. When sufficient surface has been removed, finer grit pads are then used.

It is common to polish concrete floors which have fine aggregate in the surface, such as terrazzo floors. These are often laid as tiles and final polishing is done when the bedding mortar has hardened. Polishing can be done either dry or wet; the latter is preferable since water cools the polishing pads and captures removed material, working it up to a paste that assists further polishing.

The final surface will be much affected by the hardness of the aggregate, which controls the characteristics of a polished appearance. Recently, coloured waste glass has been used to provide a distinctive brightly coloured appearance. It can also be used to form lettering, logos or symbols on a surface by casting it into confining moulds and then polishing it to expose the lettering.

By contrast, polished concrete has become popular for minimalist interiors that require a sleek and smooth finish. Colours can be added, but in the UK the choice is usually for a conservative natural colour generated from the combination of sand, cement and aggregate.

Terrazzo floor, The University of Wisconsin, USA (photo: Robin Schmidt)

Floors have to be cast or cut into areas that can handle drying shrinkage without cracking. An unintended crack will be difficult to hide by filling and is likely to become a movement joint.

Worktops and tops containing sinks are also being moulded in concrete and are available from c.£300 per linear metre for 40 mm thick sections, 60 cm wide. These can either be cast *in situ* or precast. The former can allow the formation of more elaborate profiles, and the question of weight and joining precast units is not a problem. Most manufacturers prefer to supply precast lengths up to 3 metres, so the messy casting and polishing can be done in the workshop and not the kitchen of the client. This also allows errors to be hidden from the client!

Tastes in the UK tend towards grey or natural concrete colours that provide a calmer feel in the house. In the USA, a wide variety of materials are added to concrete, such as pigments, dyes, glass, ceramics, wood, brass, aluminium, stone and pebbles.

4.2.7 Exposed concrete finishes

This section, and the book in general, does not set out to cover and illustrate all types and methods of producing exposed finishes and decorating concrete. The objective is to stimulate readers into considering and thinking about how concrete might be finished, using examples of how others have succeeded by applying their knowledge and expertise.

For a thorough grounding in finishes and guidance in exposed concrete, the reader is directed to the 'Appearance Matters' series, a collection of nine booklets by William Monks, Fred Ward and D.D. Higgins and published by the Cement and Concrete Association (and soon to be restructured, amended and reprinted).

Also recommended are two detailed texts on the subject: *Guide to Exposed Concrete Finishes* by Michael Gage (1974) and *Hard Landscape in Concrete* by Michael Gage and Maritz Vandenberg (1975).

The following sections give some examples from these references, including a limited selection of illustrations, under two headings: moulded surfaces and exposed aggregate finishes.

Moulded surface finishes

1) Moulded smooth finishes

Most smooth finishes are moulded against GRP, or steel moulds when many repeat castings are required. Smooth and sealed timber as well as plywood can be used, especially in precast work.

2) Moulded textured finishes

Fine texture can be achieved using an absorbent mould liner, the absorption causing intermittent contact between the liner and the concrete. Rope and hessian are good examples of materials that produce such texturing. These require pre-soaking in water to avoid the concrete bonding; saturated and surface dry might be the best condition.

When large panels or relatively tall shutters are planned, sealed timber can be used to produce a wood grain effect, especially if the grain is raised beforehand. If the timber is not sealed absorption over the surface will produce variation in the colour due to migration and concentration of cement paste alongside areas taking up water from the fresh concrete. Hardboard liners require oiling to prevent a bond to the concrete.

3) Moulded profiled finishes

The choice of material is governed by the size and shape of the unit being moulded. Concrete can be used to form thin sprayed materials such as GRC as well as heavy castings. Thin profiled panels are best moulded in GRP cast from a wooden master.

Shallow profiled units are ideal for casting against standard plastic or flexible mould liners held in place within wooden or steel backing moulds. Deep and intricate surface texturing, sometimes with undercuts, is best moulded with flexible rubber moulds. One-off castings suit sculptured polystyrene or polyurethane formers that are easily profiled and then simply broken away from the hardened concrete the next day. Deep smooth profiling is best moulded in steel, timber or GRP moulds.

Exposed aggregate finishes

When the aggregate is hard, well graded and well bonded into concrete it can provide a durable and aesthetically acceptable finish. Rounded, angular or crushed aggregates give different appearances. There is a vast range of colours;

however, weathering has to be considered and this may limit the range. Granite as a crushed aggregate is a better choice than oolitic limestone other than when used in semi-dry cast stone. Flint can be used in both the rounded and the knapped form.

1) Naturbetong
Single-sized material, ideally rounded, is the preferred aggregate. These finishes are formed by filling the mould with aggregate and then injecting mortar to fill the gap between the aggregate. This ensures aggregate particles touch and the exposed surface can appear without mortar. The aggregate is bonded from behind the face.

2) Ground and polished
Appearance is always of key importance. Trials should be carried out on test panels to demonstrate the effect of preliminary grinding and polishing before starting on the job proper.

3) Brush and wash
This is just a way of exposing aggregate. Compacted concrete is allowed to stiffen just enough so that brushing and washing will remove surface laitance without dislodging the aggregate. Again, a trial panel is needed alongside the main job.

4) Tooled finishes
There are four ways to mechanically produce a tooled surface: 1) needle gun finish, 2) bush-hammered finish, 3) point-tooled finish and 4) hammered rib finish. The first three relate to the shape of the head used to impact the concrete. The fourth is a technique in which the cast smooth ribs are broken off, tending to expose the aggregate.

The point-tooled finish on the Nuclear Physics Building, Oxford has weathered particularly well; the photograph was taken at least 20 years after the building was erected.

5) Abrasive blasting
This is a means of removing a surface to expose aggregate to varying depths. It is sometimes used to create visual effects, or to appear to write on the surface.

6) Felt float finish
This finish has become less popular lately due to the skill and time required to produce it. It is usually a precast operation, and produces a fine texture appearance. It can be carried

[Right] *William Mitchell casting a mural in 500mm square clay mould (photo: William Mitchell Designs)*

[Below] *Point-tooled finish on the Nuclear Physics Building, Oxford (photo: Author's collection)*

Decorative Processes and Techniques

out on floor slabs, but nowadays is more likely to be used in artistic work.

7) Aggregate seeding or broadcast (scatter) surfaces

The former is the American name for what is essentially scattering aggregate over a trowelled and wet concrete surface, then tamping the aggregate into the concrete sufficiently to ensure a bond. The surplus is then poured off.

8) Aggregate transfer

Selected aggregate is placed into suitable trays containing sand to a depth of about half the aggregate width. Close packing and appropriate aggregate selection is important, as also is the colour arrangement. The trays are then transferred into the mould. When a column is being cast, facing sheets over the trays allow transfer to a vertical position. The facing is slid vertically away as the concrete is poured. A final washing-off of the bedding sand reveals the exposed aggregate.

The technique can be used to produce a variety of objects, in addition to structures. In the example shown (see page 177) William Mitchell is casting a mural in 500 mm square clay moulds. After demoulding, part of the completed mural is coated with a pigmented resin.

4.3 Other techniques

4.3.1 Painting

It is often expected that struck concrete surfaces can be easily painted in a similar way to brick and plaster, for example. However, concrete usually requires some form of preparation since cast surfaces nearly always contain some air voidage. Surfaces formed against absorbent or permeable liners tend to result in less voidage, requiring less preparation, and are therefore more suited to painting.

Without preparation of the surface, paints often tend to make defects more noticeable, especially if thin, light-coloured paints are used. If a smooth finish is required, the best and most consistent effects are achieved by using multi-layer systems. Paint can be used to colour a textured concrete surface, but depending on the exposure a stain may be a better option.

Whatever the finish and colouring treatment proposed, several steps can be taken to maximise the likelihood of success:

1   Shuttering must be free of steps between adjacent panels or features used to conceal abutting pours.

2   Any significant holes or voids must be filled. This can be done with polymer modified mortar or proprietary filling systems.

3   It must be made clear from the outset that the concrete is to be painted, and sample panels cast to confirm a suitable surface can be produced.

4   A treated panel, produced on site, should be set up as the approved sample.

5   Details of the paint and application technique used must be available on site. It is wise to leave an area of the concrete sample clear of the treatment in case any further trials are needed.

Whatever paint is used, the full preparation procedures recommended by the supplier should be followed. These may involve conditioning the surface alkalinity, accounting for moisture in the concrete, and cleaning up loose or dusty surfaces.

Paints are often considered as simply a uniform coating used to cover up an otherwise defective or repaired surface. That is often the case, but they can also be used to enliven an otherwise dull surface or structure. Power plant cooling towers are mostly drab, although their shape is interesting. In Germany a cooling tower has been painted with a complete map of the world, and colourful internal paints for concrete have been around for some time.

Where a durable painted surface is required, such as on the outside of buildings (especially in aggressive environments), paints formulated using a silicate base have been shown to provide the best long-term protection. Some proprietary paint treatments, based on silicate binders, have been around for 100 years or so and still provide a satisfactory finish.

The Deep Aquarium, Hull by Sir Terry Farrell & Partners is an example of a building (see overleaf) with a relatively low pigmentation colour wash that acts as a penetrating stain to equalise colour variation whilst maintaining the texture. These products are a combination of potassium silicate, inorganic mineral fillers (mainly feldspar) and inorganic earth oxides.

When internal paint finishes in strong colours are required, a liquid silicate primer coat is often used to

Decorative Processes and Techniques

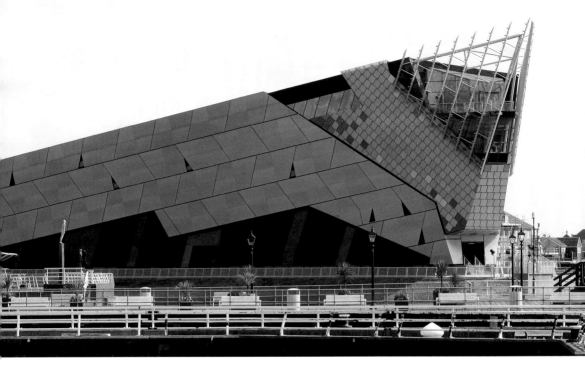

The Deep Aquarium, Hull; Sir Terry Farrell & Partners
(photo: © Gordon Ball/Dreamstime.com)

Decorative and Innovative use of Concrete

reduce surface porosity as well as acting as a bridging coat
to cover fine (c.1 mm) cracks before the second or final
coat is applied. Such a system has been used in Manchester
Airport.

Most of these colourful silicate paint systems come
from Europe and have been used there for some time. An
example of one such system, used in Iceland, the Baltic and
Nordic countries for over 30 years. The system has been
shown to be resistant to extreme climates and is ideal for
marine environments. It uses an alkylalkoxysilane primer
that acts as a liquid water repellent but allows water vapour
to pass through the coating, preventing vapour pressure
build-up that might otherwise blister the coating. This is
especially appropriate on surfaces subjected to heat from
sunlight or when conditions vary across the paint layer. The
system is a combination of hydraulic lime, white cement
and inorganic fillers.

Colourful silicate paint. St. Benno Gymnasium, Dresden School;
Benisch Architects (photo: Christian Kandzia)

*The Royal Opera House, Covent Garden, London
(photos: Phil Henson, http://www.flickr.com/
photos/24583241@N04)*

Decorative and Innovative use of Concrete

4.3.2 Thick paint/thin render systems

Where a solid coating is required, a thicker system is needed (e.g. as used on the exterior as well as the interior of the Royal Opera House). Thick coat systems designed to bridge cracks, especially live ones, tend to be polymer modified and contain glassfibre mesh reinforcement.

Polymer modified mesh reinforced systems are usually installed by approved companies, listed and monitored by the material formulators and suppliers. The systems are now only offered on vertical surfaces with horizontal ledges such as window sills and parapets formed with other materials or capping products. These coatings have been used successfully on housing developments and commercial buildings but there is no reason, other than the high cost, for using such low maintenance systems on private housing.

4.3.3 Renders or overlays

Cement-based renders have been in use for many years, and provided the correct sand/cement ratio is used and the sand grading is appropriate they give good performance. Often one sees cracked render and spalling due to either movement of the substrate or excessive shrinkage of the render, or both.

Proprietary renders are now available that combine hydraulic lime, white cement and mineral fibre fillers. These can accommodate a high degree of movement whilst preventing or distributing surface cracking or crazing. They come pre-bagged and can be bought with pigments added. They can be used with reinforcing mesh over critical areas subject to movement.

Another use of such renders is to apply them over a grid of timber battens. When these are removed, the render takes on the appearance of stone masonry construction.

4.3.4 Decorative saw cutting

Flat concrete surfaces are often enlivened by cutting lines into the surface to form borders or to profile shapes, patterns or the outline of animals or figures, etc. A common use is to install a company logo in the floor of an entrance lobby or shop.

The lines form boundary areas that can then be chemically stained with acid-based oxide stains or dyes applied to the surface. An unusual effect can be achieved by cutting narrow grooves into a concrete surface that has

[Above] *Dacorum Leisure Centre, Hemel Hempstead. Faulkner Brown. (photo: Martine Hamilton Knight)*

[Right] *Premiere Place, Docklands, London. Chassay and Last. (photo: David Groom)*

Decorative and Innovative use of Concrete

already been stained or is of a suitable natural colour. The cut lines are then continued to form a structure such as the branches of a fern. All the cut lines are then cleaned of dust with a vacuum cleaner and acid stain colouring is fed into them with an eyedropper, taking care not to spill the stain over the edges of the cuts.

Before the liquid has been taken up by absorption a fine nozzle on a hand air tool is used to blow the stain from the cut lines, at right angles to the line, making a splatter pattern. For a convincing appearance it is important the colours of the base and stain are compatible. Stains and dyes are described in section 4.1.

Blades have been specially developed for this type of surface cutting. They have spiral flutes on the sides of the rim that clear cut material from the side, both preventing undercuts and providing a crisp cut line.

4.3.5 Engraving

Engraving of concrete is a form of saw cutting on a macro scale. It can be used both with old concrete in need of some improvement and with new surfaces which are cast ready to be cut.

The engraving machine can be set to cut various shapes such as circles or lines. The surface is cut away to form dummy mortar joints rather like the effect obtained by stencil patterns. The joint width can be adjusted between 10 and 25 mm but the depth of cut is only a few millimetres. Low depth prolongs the life of the blade (to provide a total run of c.9000 metres). Cutting can be wet or dry, but dry cutting with a vacuum attachment is preferable.

Some surfaces may not suit the process, especially if they are voided or spalled. In these cases a thin repair topping cut or formed by stencil patterning would be a better choice.

Before the concrete is cut it is usually coloured by one of a range of methods, including chemical stains, colour sealers or polymer (acrylic) based sealers. Circle cuts are preferred because once the machine has been connected to the pivot point, forming the centre of the circle, the parallel radii can be easily cut by clamping the machine at increments out from the centre of the pivot bar. Cross cuts are formed by setting the machine round 90°. When combining machine and hand cuts with colour, the patterning effects can be spectacular.

Metal sprayed on concrete (photo: William Mitchell Designs)

4.3.6 Metal sprayed concrete

Metal spraying has been around in the engineering industry for some time. A spray gun is essentially a thin hot flame into which a metal rod or wire is fed continuously, allowing the vaporised metal to be expelled by the flame. Once it hits a concrete surface the hot metal vapour is instantly cooled and the metal forms a continuous skin. Application is somewhat akin to paint spraying.

Concrete can be coated with zinc and probably other metals such as copper and aluminium. If the concrete surface is dry there is less chance of a reaction between the metal and the alkali in the concrete. I have practical knowledge of spraying zinc onto concrete to coat 'dog-bone' shaped concrete sticks for casting into structures as crack gauges. When the resistance of the film rises, the metal skin is being stressed and when the resistance is infinite the film has fractured, indicating significant movement.

Concrete surfaces are best acid washed and then thoroughly dried before spray coating to ensure any loose laitance, etc. has been removed.

4.4 Glassfibre reinforced concrete

Glassfibre reinforced concrete (GRC), or glassfibre reinforced cement as it used to be known, is a composite material formed from cement, fine sand and glassfibre reinforcement. Early examples had higher cement content than current formulations and initially some products had a minimal amount of sand. This led to significant shrinkage and associated cracking.

Current applications contain equal amounts of sand and cement, albeit fine sand, 2 mm to dust and other additions such as microsilica or metakaolin that react with free lime released from the hydration of the cement. It is also now common to include polymer dispersions to improve the fibre/cement interface characteristics, reducing the erstwhile tendency for the material to become brittle with time. In the UK the material has not sustained the product output seen in Europe and elsewhere, despite the first commercial fibre being manufactured here and GRC production being controlled by a licence agreement with Pilkington Glass.

A dramatic early application was a temporary octagonal roof structure at the 1977 Stuttgart Garden Show. Eight curved sections were jointed together, and spiral reinforcement was encased to form a curved supporting rib. Testing later found the reinforcement in the ribs was not necessary. The structure was 31 metres in span and the GRC thickness was 10 mm. With its shape and apparent simplicity, the structure represents a decorative and innovative use of GRC.

Glassfibre reinforced concrete can be sprayed onto a mould (as was the Stuttgart roof) or cast. The product has many applications but has become particularly popular for renovation projects. The Shepard Hall, New York City (www.ccny.cuny.edu/aboutus/campus/shepard00.htm), 1907 was built from a combination of stone, terracotta and steel, and by the mid-1980s major restoration was needed. The possibility of using new terracotta and cast stone was rejected due to their poor past performance in the polluted New York environment, and metakaolin modified GRC was used instead. Over a 12-year period 23,000 m^2 of terracotta has been replaced with GRC. Phase 1 was 12,000 units, Phase 2 another 6200 units, Phase 3 a further 11,500 units and Phase 4 around 8500 units. The units include clerestory windows, copings, finials, mouldings, quoins, flat panels and sculptures. Before GRC was chosen a vigorous test programme eliminated other materials that discoloured or yellowed and showed lime bloom during acid rain conditions. Cem-FIL® Star GFRC was finally chosen; this has a combination of cement and a pozzolanic cementitious addition that can convert free lime (which deposits lime bloom) into calcium silicate.

Glassfibre reinforced concrete can be made to resemble stone by facing a mix during the casting process with reconstructed or cast stone. A facing is first laid in the mould and GRC spray applied to form a monolithic bond between the two. Plain GRC may be the preferred option, moulded to include profile shapes. A recent renovation programme of Jaguar dealerships in the UK included curved panels for entrance porticos in GRC.

Glassfibre reinforced concrete in use is lightweight because it is created in thin sections. Acoustic barriers usually require mass to attenuate noise but lighter weight versions can be formed in GRC which reflect noise and suppress transmission by the inclusion of a lightweight core. There are thus numerous applications of GRC noise barriers.

In theme parks, holiday camps and elsewhere GRC has been used to form artificial rock structures. In the hands of skilled and knowledgeable installers these can be made to look surprisingly realistic. If rocks can be cast it is relatively easy to spray or cast wall sculptures provided a suitable detailed mould has been taken from a sculpture's master. In fact the master, if it is clay, can be coated with GRC, thus avoiding the need to take a moulding.

Glassfibre reinforced concrete is particularly popular in the Middle East, where it has been used to form numerous additions to palaces and municipal and private buildings.

4.5 Cast stone

Cast stone is manufactured in two ways. In the first a semi-dry mix is compacted into the mould to remove voids and the surface closed against the mould. Demoulding follows immediately afterwards. The durability of the material produced depends on the grading of the aggregate used and the amount of compaction achieved.

In the second, wet cast, the procedure is more similar to that for conventional concrete, involving casting into a mould, vibration, and then demoulding the next day. The density and strength of wet cast are usually higher than semi-dry and it weathers better due to the surface being denser and more consistently compacted. Semi-dry is usually a better match to natural stone since the surface is more open and granular, but it has high water absorption that tends to promote organic growth. Wet cast requires an acid wash to remove surface laitance, and the appearance is more of a ground stone surface. It has lower water absorption and less risk of lime bloom, and is more resistant to mould growth.

Chapter 3 includes an account of the history of cast stone and how it was and is made. The earliest attempts at

[Above] *Large curved pre-cast reconstructed stone panels. The Edinburgh Conference Centre (photo: Hamish Irvine)*

[Following page] Detail from *The Edinburgh Conference Centre (photo: Kristie De Garis)*

producing concrete artefacts involved creating a kind of cast stone, since the aim was to provide an alternative product within a market that until then had been dominated by natural stone, Coade stone, Ransome's patent stone, and so on. Indeed, Joseph Aspdin named his invention 'Portland cement' because he said it resembled Portland stone.

By the 1920s growth in the cast stone business had led to the formation of the British Precast Concrete Federation. At that time the Café Royal was clad in semi-dry cast stone which appears to have weathered well, apparently confirming the potential of semi-dry material. Unsatisfactory finishes are, in my experience, often caused by failure to use the correct aggregate grading, resulting in the surface not being dense and compacted enough.

Manufactured correctly, cast stone can be moulded into contemporary masonry products and used for cladding and paving, as well as other intriguing applications. A new facing medium that simulates flint walling has been developed. The blocks are built into the outer walls of a cavity construction and the joints and dummy joints filled, providing an authentic alternative to natural flint walling.

On roofs in the Cotswolds it is traditional to use stone slates that decrease in size towards the ridge. These are now simulated in concrete by casting against a flexible textured mould. The mottled colour is achieved by incomplete mixing of pigment into the mix or by combining two concrete mixes of different colours and then partially mixing them before casting the units.

Practically every element of the external elevation of a building can be produced in cast stone as with the Edinburgh Conference Centre, and internally it can be used for stairs and flooring.

4.6 Polymer concrete

A version of polymer concrete, a mixture of a resin and sand, is now being used to cast kitchen worktops and sink surrounds. Often the polymer is methyl methacrylate (Perspex®). This type of concrete can have a high sand content, up to about 96%, and forms a surface similar to marble.

Minerelle™ is a combination of natural mineral sand that forms a smooth, renewable surface and allows inconspicuous jointing and edging. Zodiaq® contains 96%

Decorative and Innovative use of Concrete

The Bude Light, Bude, Cornwall; Carole Vincent (photo: Chris Williams)

quartz sand and 4% resin and is used for vanity tops and worktops. It has a hardness of 7 on Mohr's scale, the same as quartz. Silestone® combines 93% natural quartz with resin and pigments that are pressed to form and assist bond. It has high stain resistance and low water absorption.

4.7 Other processes

4.7.1 Artistic uses of concrete

Carole Vincent

Carole Vincent was born in Crediton, Devon, the daughter of a builder. She started in the family concrete works at the age of nine casting blocks. After studying at the Bath Academy of Art she set up her own workshop and studio in Boscastle, where she continues her work and runs summer classes.

Her work with pigmented concrete has won wide acclaim. She works on commissions for both private gardens and public locations and takes great care with designing her creations to suit the specific environment. There is little that resembles the preconceived notion of concrete in Carole's work.

Bright vibrant colours, clean lines and profiles and highly polished surfaces characterise her work. Her success stems from her understanding of the material and from the ways she has developed to exploit the benefits of self-compacting concrete utilising superplasticisers, viscosity agents, additions of accurately graded fines and cementitious additions such as metakaolin.

Carole works by moulding her master form in clay and from that taking a plaster or GRP moulding, often fragmented into many pieces to avoid undercuts locking the mould to the casting. Carole's knowledge and the meticulous way she has developed her art and understanding of concrete to suit her work are obvious from the accounts of trials she has published. Her book *Concrete Works* (2003) gives an account of some of her main works, from relatively small to much larger. The latter two include fibre optic lightning.

Carole won a Bronze Medal at the 2001 RHS Chelsea Flower Show for the design and construction of the artefacts in the Blue Circle Garden, gaining her further and lasting recognition. All the items were cast at her studio in Boscastle.

In addition to her still-life structures, Carole has also produced fine sculptures in a more subdued, ground or polished exposed aggregate coloured concrete. These have a flowing characteristic that imbues great feeling into the minimalist concrete form.

In *Concrete Works*, Carole lists the following commissions and public works:

1982	Reunion Commission; Miss A.M. Shaw
1984	One and All Commission; W.E. Chivers Ltd, for Old Vicarage Place, St Austell, Cornwall
1985	The Family Location; The Surgery, Boscastle, Cornwall
1988	The Armada Dial Commission; Plymouth City Council, for the city centre
1989	The Devon Pedestrians Commission; Devon County Council, for Exeter, Plymouth, Torquay and Barnstaple
1990–91	Quartet Commission; Marina Rainey for Oldhay, Launceston, Cornwall, and InterCity for The Royal Scottish Academy of Music & Drama
1993	The Peacock Bowl at Chelsea Flower Show
1993	The Buskers Commission; Safeway Stores, for Cage Yard, Reigate, Surrey
1994	Colloquy Two Commission; Ssang Yang Cement Ltd, for their offices in Singapore
1995	Colloquy Two 3/5 for David Bows
1996	Les Jongleurs Commission; Jersey Public Sculpture Trust, St Helier, Jersey
1997	Atlantis Commission; Private client, Jersey
1998–99	The Red Carpet Commission; The Edinburgh Festival Society, for The Hub, Highland Tolbooth
2000	The Bude Light 2000 Commission; Rude-Stratton Town Council, for the Castle, Rude, Cornwall
2001	The Blue Circle Garden Sponsor; Blue Circle Industries plc, at RHS Chelsea Flower Show
2003–04	Concrete Jungle Proposal for RHS Chelsea Flower Show

Carole Vincent's output probably represents the most consistent and inspiring attempt at lifting concrete out of the 'grey and dull' concept held in the mind of the general public.

Janette Ireland

One could say that Janette Ireland achieves her style of decorative concrete by covering it up. The surfaces are filled with pebbles to form rows, patterns, pictures, etc. This could

Decorative and Innovative use of Concrete

[Above] *The design is worked out and drawn (photo: Janette Ireland)*

[Left and below] *A mould is constructed (photo: Janette Ireland)*

[Below left] *The aggregate is covered with a shrink-compensated grout (photo: Janette Ireland)*

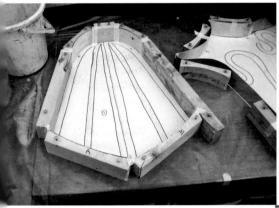

Decorative Processes and Techniques

[Above] *The aggregate is covered with a shrink-compensated grout (photo: Janette Ireland)*

[Below] *When the grout is firm concrete is poured over (photo: Janette Ireland)*

196

[Above] *The mould is stripped (photo: Janette Ireland)*

[Left] *Sand is brushed off (photo: Janette Ireland)*

[Below] *The work is assembled like pieces of a jigsaw (photo: Janette Ireland)*

[Above] *Individual cast pieces fit to form the finished work
(photo: Janette Ireland)*

[Opposite] *A completed project (photo: Janette Ireland)*

Decorative and Innovative use of Concrete

be considered similar to the aggregate transfer discussed in section 4.2.7. However, the difference is that Janette constructs her works by selecting each piece of aggregate individually and placing it carefully in order to construct a picture or pattern, whereas normal aggregate transfer produces plain aggregate texturing.

The design is worked out and drawn. It is then transferred to mirror an image that is covered in perspex. A mould is then constructed over the drawing using wood, polystyrene or strips of aluminium. This enables sections of the overall design, that might be for a pathway or patio area, to be cast for assembly later on site.

Janette uses rounded as well as elongated pebbles of various colours to form her designs. These are graded ready for her to select. The chosen pebbles are built up in the mould following the underlying pattern and set in dry sand to a depth of 5–8 mm. Once a section has been completed the aggregate is covered with a shrink-compensated grout to a depth of 25 mm to lock the particles together. When the grout is firm, but not hard, concrete is poured over the grout.

After a 24 hour curing period, the mould is stripped, the unit turned over and the surface sand brushed off. Each unit is then individually wrapped in polythene and cured for 28 days. The work is then assembled, like pieces of a jigsaw, to create the final form, in this case a garden pathway.

Great care and attention has to be given to the detail and sizing of the pebbles to ensure the individual cast pieces fit to form the finished work – a time-consuming process. The completed works are stunning, and have the advantage of changing appearance under different weather conditions.

Taking these illustrated examples, one can appreciate the time involved in composing the picture (by selecting each piece of naturally rounded aggregate individually) and then carrying out the moulding sequence described above.

Federico Assler

Federico was born in 1929 in Chile, started his working life 'working with tools for making furniture'. He began studying architecture but left after two years to start painting and attend art school (see http://www.federicoassler.cl/).

Painting became his mainstay. He had exhibitions in Mexico, the USA, Argentina and his native Chile, and designed scenery for the National Ballet, as well as a special children's playground. He became a Director of the

Contemporary Arts Museum of Santiago. In 1971 he began his work sculpturing in concrete.

Federico is noted for free-standing sculptures and large sculptural wall forms. In 1991 he spent a year in England, made possible by the combined sponsorship of the British Cement Association and The Building Centre as part of the Creativity in Concrete programme. An exhibition of photographs of his work was mounted, which included illustrated presentations by the artist himself.

Edward Lazenby

Edward (Ed) Lazenby is one of a few 'hands-on' exponents of decorative concrete who can produce quite amazing effects and features out of concrete. He began by producing patterned-imprinted monotone concrete, then developed a wide array of coloured, stained and tinted effects.

When the Princess Diana Memorial Playground in Kensington was designed by Mel Chantry, concrete proved to be the ideal medium for its construction and Ed Lazenby worked on site with the designer/artist and the concrete contractor. Freehand forming was achieved with concrete over a four-day working period, avoiding the appearance of joints and contrasting colours between the pours.

The inspiration was taken from beaches in Lanzarote. The idea was to mimic the sweeping curves of sand, while also incorporating some petrified wood simulations into the concrete. The clay-like working characteristics of plastic concrete allowed it to be moulded and carved into the final profile. Objects which have been created include pebbles, coral, sea-worn glass, footprints, fish, leaves and starfish.

Plain paving is not the business of Ed Lazenby. Techniques he has developed can produce waves in the surface, as seen at Port Seton Promenade in East Lothian, Scotland, as well as seahorses and leaves. At Possil Park, Glasgow, latex-applied stencils were used to form 3500 letters; the unprotected area was shot-blasted to remove the surface, exposing a silica-rich under layer.

Sometimes existing concrete surfaces are used as the base and chemical stains applied to cut lines and patterns in the concrete, forming translucent colouring effects. This technique has been used to produce, for example, a map of Europe, a forest trail including lion and gazelle prints, maps of the night sky and large areas of text and mathematical equations inscribed into the surface of concrete.

The Group of School Children, Co. Westmeath,
Ireland; Imogen Stuart (photo: Imogen Stuart)

Cillian Rogers

Cillian Rogers had an eye for the comical side of sculpture and in his *The Time of Day* he invites the viewer to join in by sitting next to the two men, one with a bicycle and the other already seated on the bench. A bicycle was encased in concrete to form the feature, which also includes reinforcing bars and expanded mesh. The sculpture is in Easkey, Co. Sligo and is a popular item with the locals.

Imogen Stuart

An older example of Irish concrete sculpture is found in Tyrrell's Pass, Co. Westmeath. It was made in 1970 and given the name *The Group of School Children*. This study was modelled in clay and a plaster mould taken, into which concrete was placed. It is thought a reaction between the plaster mould and the concrete produced the fine finish. Rapid-hardening cement concrete over a steel armature was used, and it has evidently matured well.

Ivan Coghill

Ivan Coghill has used the art of pargeting to form sculptured exterior and internal wall features. Pargeting dates back four centuries and is found on the exterior of buildings in the eastern counties of England. It is normally carried out using plaster. The plaster mix originally contained a variety of ingredients such as horsehair, manure and wood shavings, which today might be replaced with fibre, pigment and lightweight aggregate.

Pargeting is primarily a form of wet-working to generate the relief patterning. Sometimes casts or precast items may be incorporated, and the relief may be carved after the medium has hardened. Dramatic effects can be produced when the parget, formed into striking representations, is applied to outside walls. It can also introduce surprising effects on interior walls.

More conventional plaster pargeting is carried out by Anna Kettle.

Rachel Whiteread

The winner of the Turner Prize in 1993 was *House*, a hollow gunite casting of the interior of the last remaining house of the 19th-century terraced row in Grove Road in Tower Hamlets, East London.

Here the house was the mould and concrete was sprayed inside before the house was demolished. The

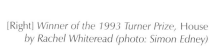

[Right] *Winner of the 1993 Turner Prize,* House
by Rachel Whiteread (photo: Simon Edney)

[Below] *Concrete luggage sculpture, Liverpool
by John King (photo: Christopher Hooper)*

Decorative and Innovative use of Concrete

work was sponsored by Tarmac Structural Repairs, The
Arts Council, the London Arts Board, The Henry Moore
Foundation and Artangel Patrons.

The finished structure eventually gained two awards:
the Turner Prize as mentioned (being a cheque for £20,000)
and one from the K Foundation. This second award, of
£40,000, was for the worst work of art. Rachel initially
refused to accept the cash prize and only when the K
Foundation threatened to burn the money on the steps of
the Tate Gallery did she accept it, passing it on to charity.

The structure ended up being demolished in front of the
press in January 1994.

John King

Concrete luggage was cast and set up in Liverpool at the
junction of Mount Street and Hope Street. Acid stains were

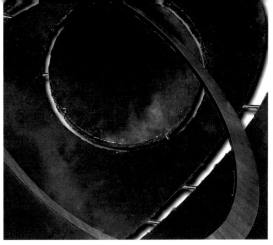

[Left and above] Terracotta wall sculptures; David Undery (photos: David Undery)

[Below] Installation of a grey composition; David Undery (photo: David Undery)

used to provide the colour to the cast items. The artefacts form a novel feature for children.

David Undery

David Undery became acquainted with concrete whilst working as a site labourer in the summer holiday breaks while he was doing his fine art degree. This changed his conception of concrete as being a cold, bland substance. He now sees it as a medium suitable for producing his highly visual, lustrous and tactile hand-crafted relief sculptures.

Examples of his work are hung in private homes. His work can also be found in a number of restaurants in the West End of London.

David's work involves introducing texture and pigment into the concrete. Acid staining is also employed, as well as inlays of metals, including gold leaf, and of wood. Concrete has a reputation for being heavy but can be made relatively lightweight by incorporating a voided backing and lightweight aggregates or a foamed structure.

David has developed a way of making his work strong and slimline, and light enough to be hung on a picture hook. Lightweight reinforcement is also incorporated.

Peter Kennedy

In a similar way to David Undery, Peter Kennedy of the London-based Petr Weigl studio has produced concrete wall art. He offers this under a range of four forms: Contour, Erosion, Fossil and Bohemian Legacy.

4.7.2 Printing on concrete

Until recently images such as those in newspapers were formed from a series of minute black dots in rows, of varying diameter. Over dark areas the dots had a large diameter and adjacent dots joined together. Grey areas would have medium diameter dots with white areas between, and white areas no dots. Lines could be formed by the dots tracing the required location of the line.

A similar technique has now been used to transfer an image onto concrete. The dots are areas of surface erosion that cause a light-coloured surface to be removed, exposing a darker aggregate beneath.

A close-up of the effect, showing the individual dots, is illustrated. It has an appearance like that of a picture in a newspaper or a magnified image on a computer screen. The

Photo-engraved concrete (photo: Sergio Cobos)

etching procedure is as follows. A special paper is treated with a retarder in a representation of the image; in other words, the paper has the retarder applied in a series of dots the diameter and quantity of which reflect the picture form. This sheet is then placed in a horizontal mould and the concrete poured over. It is prudent to use a self-compacting or superplasticised concrete to avoid disfigurement of the image as it transfers to the concrete. Also, to ensure good compaction the concrete must have a continuously graded particle size of sand and aggregate.

The Pieri's Serilith system is a system where a design is produced on a polystyrene sheet as a grid of dots of retarder. The sheet is then placed in the base of the mould and used once to transfer the retarder to the concrete. After two days of curing the unit is demoulded and the face washed to remove the retarded surface.

The effect can be quite stunning and detailed. Dark aggregate, pale sand and white cement seem to provide optimum results. This technique is ideal for many applications.

T his chapter provides examples of a range of innovative concrete textures, finishes and profiles, found mainly on buildings, from across the world. The objective is to show pictorially forms that might not otherwise have been considered possible or practical.

The range is diverse, from church wall relief panels by William Mitchell viewed at close range (exhibiting subtle and ingenious fine texture), through the sinuous building forms and structures of Santiago Calatrava, to the enclosing private spaces encountered in the work of Tadao Ando.

With the exception of some notable sculptures, one nearly 100 years old, practically all the examples are modern. It is hoped that together they succeed in demonstrating that the possibilities of concrete are restricted only by the imagination of the user.

I make no apology for starting with a section on the work of William Mitchell. William Mitchell is not as well known now as during the 1960s to 1990s when he was most active. His style is also not in the current minimalist vogue. However, his range of work, his abilities in forming, texturing and colouring, and his innovative methods of preparation make him one of the foremost exponents of work in concrete.

Recent works in concrete by some notable individuals

5.1 William Mitchell

William George Mitchell is something of a 'one-off' in terms of both the type of work he has produced and the way he became involved in decorative concrete. He was born in 1925, and spent a large part of his early years in hospital and convalescence homes suffering a serious illness. In 1938 he took up an apprenticeship to a firm of decorators. Following War Service in the Royal Navy, he spent two years decorating NAFFI clubs and canteens, painting scenes and panoramic views, all over the world, from Freetown in West Africa to Corsham in Wiltshire.

Following this, an application to Portsmouth College of Art to pursue an N.D.D. (National Diploma in Design) course was turned down on education grounds so he went away and sold insurance and returned a year later with the cash for the course in a small cardboard suitcase.

His studies were aimed at a career in restoration of classical buildings, an area in which there was much need following the damage during the war. Mitchell went on to the Royal College of Art school of woods, metals and plastics and a fourth-year postgraduate award in the British School in Rome, obtaining a silver medal.

On returning from Rome, Mitchell took up an appointment with London County Council Architects Department, working on designs for housing estates. This gave him the opportunity to work alongside builders, architects and engineers on a vast range of different projects. Since then his 55 working years have embraced the use of concrete in a way that is characteristic only of William Mitchell.

The first illustration is a typical example of his work: a wall tablet, one of eight used as gargoyle fountains in the water garden designed by Frederick Gibberd and commissioned by him for Harlow New Town.

Chapter 3 mentioned Mitchell's use of Faircrete and the way he has learnt to work the thixotropic concrete into forms, retaining the profile precisely as the concrete stiffens and hardens. He is probably best known for 14 Faircrete wall tablets (c.1.5 × 2 metres) depicting the Stations of the Cross (page 86), housed in alcoves in the internal walls of Clifton Catholic Cathedral in Bristol.

Working fresh concrete into a required form obviously takes time – but that time is limited to about 1.5 hours. Mitchell worked by transferring a charcoal outline from a drawing onto the prepared concrete surface and then peeling the drawing away. . The outline, texture and profile then has to be applied to the concrete in such a way as

[Above] *William Mitchell; 'Watergardens 2' Harlow New Town, 1973 (photo: William Mitchell Designs)*

[Right] *William Mitchell; Harlow Fountain Court (photo: William Mitchell Designs)*

[Left] *William Mitchell; Moulding the hand symbolising Christ crucified, Bristol Cathedral (photo: William Mitchell Designs)*

[Below] *William Mitchell; The hand symbolising Christ crucified, Bristol Cathedral (photo: William Mitchell Designs)*

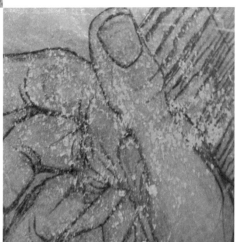

to depict the intended meaning. William Mitchell has developed an ability to make plain concrete into faces, gestures, objects and feelings that truly convey a meaning.

Some of William Mitchell's work was incorporated into Harrods and the Paris Ritz. However, Mitchell does not confine his work to the eyes and presence of the few; he also has it displayed in towns and on the most utilitarian of buildings.

There has been a resurgence of interest in decorative paving of late, block paving and imprinted concrete now being common and widespread. Back in the 1960s, however, it was not as readily available, so when Queen Alexandra's Royal Army Corps in Aldershot wanted a

[Opposite page] *William Mitchell; Humanities Building Manchester University (photo: William Mitchell Designs)*

[Left] *William Mitchell; Hampstead Civic Centre (photo: William Mitchell Designs)*

[Below] *William Mitchell; Queen Alexandra's Royal Army Nursing Corps – completed courtyard (photo: William Mitchell Designs)*

[Below left] *William Mitchell; Queen Alexandra's Royal Army Nursing Corps – inauguration day (photo: William Mitchell Designs)*

[Above and left] William Mitchell Retaining Wall, Kidderminster Ring Road
(photo: William Mitchell Designs)

centrepiece in the form of a foundation stone feature to the courtyard outside a new building they had to commission the work. Designed by Mitchell, the decorative paving linked the central stone to the structure, symbolising the unity of the Corps.

The centre stone was convex in shape to prevent ponding in the large flat surface. Such considerations are too often neglected today when paved areas of any significant size are laid.

A section of the Kidderminster ring road passes alongside a tall retaining wall faced with a series of repeating panels designed by William Mitchell. Tree growth now camouflages much of the elevation but it was originally constructed to include a small waterfall. Perhaps other considerations meant the waterfall had to be abandoned. The panels have weathered well and retain the crisp features moulded into the surface.

Although sometimes large panels were cast incorporating his designs, Mitchell often formed the overall design by using a series of smaller panels fitted together, with the design flowing from one panel to the next.

Perhaps the most graphic examples of William Mitchell's work were those made for the Cement and Concrete Association, who commissioned a group of figures to illustrate as many methods of producing texture, finish and colour as possible. Mitchell chose to execute a version of the Anglo-Saxon fertility rite known as the Corn King and Spring Queen. Each figure was a different finish and colour yet they held together as a sculptural group.

5.2 Antoine Predock

Antoine Predock's Zuber House, built between 1987 and 1989, is located in desert surroundings in Phoenix, Arizona.

Decorative and Innovative use of Concrete

William Mitchell; The Corn King & Spring Queen
(photo: William Mitchell Designs)

The house is horizontal like a Prairie house and reflects the desert colours and shapes, similar to Frank Lloyd Wright's Taliesin West house nearby.

Predock's interest in Indian civilisations meant that in his earlier years he liked to make use of adobe. This has given way to concrete mainly due to his houses becoming bigger and more complex, requiring the greater structural capacity.

A series of houses were designed and built during the 1980s, starting with the Fuller House, but Predock's works also include other forms such as museums, theatres, libraries and a clinic. He lives in New Mexico and his style is highly appropriate for the local terrain. (See www.predock.com/ZuberHouse/Zuber%20House.html).

5.3 Augustin Hernandes

La Casa en El Aire, 1989 is a unique exercise in technical skill to accommodate a steep sloping terrain. Concrete twin walls rise from the slope and support the house, that penetrates the concrete circular openings as a rhomboid prism. The exercise appears visually interesting and invites further inspection to discover how the interior has been detailed. (See www.epdlp.com/arquitecto.php?id=3250)

5.4 Luigi Snozzi

Casa Bernasconi, Carona, 1988-89 by Luigi Snozzi has taken the Corbusian model and attempted to refine it into something more sympathetic to the Swiss terrain, but the form still appears built to machine aesthetics. His dream apparently is 'to build a house that is hardly seen at all, reducing everything to the indispensable'.

Snozzi works in the Italian-speaking region of Ticino, between Milan in the south and Berne in the north, where the Ticinese Tendenza have been left to perpetuate elements of architecture that have been dismissed elsewhere.

5.5 Mario Botta

Mario Botta is another of the Ticinese Tendenza, like Luigi Snozzi. Botta has moved on from Snozzi by combining influences from other sources as well as his own imagination.

Casa Bianda at Losone, Ticino, 1987-89 has been described as a cake with a few slices cut out. It rises up as a circular tower in banded, alternating blockwork, a feature of a number of Botta's buildings. Whatever is thought of the appearance of this house, it is undeniably a fascinating example of the use of what is otherwise naked blockwork.

Casa Breganzona, Ticino, 1984-88 is again a three-storey house but has opened up the circular form and intersected this with rectangular wings, giving a more complex volumetric composition. The house again uses concrete blocks which alternate up the exterior walls with horizontal strips of silicon-glazed bands, as found on many of Botta's buildings.

Botta has said: 'I love churches because they make you feel you are the protagonist. In fact you must be able to enter a church and feel that you are at the centre of the world'. The Church of Saint John the Baptist is circular, a shape which must assist those who enter to experience Botta's protagonist feeling. The banded style, something of his trademark, is used again over the walls and across the floor. The light flooding in through the inclined glass roof also appears in a banded form, intersecting the walls. According to Botta:

> Light governs space; without light space does not exist. Natural light brings plastic forms to life, shapes the surfaces of materials, controls and balances geometric lines. The space generated by light is the soul of the art of architecture.

The Morbio Inferiore School 1972, was the result of a competion entry and the first large scale project undertaken by Botta, located in open ground, surrounded by hills and fronted with a landscaped space including the protruding sculptured figure elements by Peter Selmoni.

There are eight repeated modular blocks with the classrooms on the first floor; ground floor being used as common areas and along the top or second floor, a skylit central passage runs the length of the building creating emphasing the linearity of the complex.

Evidently, the pupils dislike the building with the heavy, and to them, depressing mass of concrete forms that depict the structure. However the starkness together

Mario Botta, Morbio Inferiore, 1977 (photo: Lorenzo Bianda)

220

[Below and opposite] *Mario Botta, Morbio Inferiore, 1977 (photo: Alo Zanetta)*

[Following page] *The pupils apparently dislike the stark nature of the exposed concrete (photo: Alo Zanetta)*

Concrete Applications from Around the World

with the reflective nature of the concrete presents a light and beautiful space that may have been more suited as a monastery.

In this early building we can see the form reduced to orderly essentials akin to the designs of Kahn and Le Corbusier (Botta worked for both). The overall layout is such as to make the functions apparent with repeated porches, light wells and vertical circulation areas, far different to the curved binary coloured buildings that were to follow. The question is how such plain concrete megastructures will weather, certainly not likely to retain the uniform white appearance as shown in these early photographs unless the surface is treated.

5.6 Takasaki Masaharu

Zero Cosmology, Kagoshima, Kyushu, Japan, 1989–91 is a house made entirely of cast concrete. Uncoated both externally and internally, it shows an uncompromising use of the material. Earlier uses of concrete had tried to cover up board markings, tie holes, and so on, but this post-war example of brutalism rejects the work of fettling the surface to a pure form and respects the natural inclusions in the surface.

The house is on a tight floor plan in Kagoshima, on the largest of the southern islands, and is surrounded by unattractive buildings. The living room is perhaps the most outstanding of many features of this house. It is an ovoid concrete volume supported on cross-bracing of concrete beams that suspend the room over a pool of water, only seen from the outside. The room is entered through a circular opening and is without any furniture other than cupboards fitted below the seating around the perimeter. The rest of the house – bedrooms, kitchen and so on – is located in a more conventional form on a rectangular building to the side of the living room.

5.7 Eisaku Ushida and Kathryn Findlay

The Truss Wall House is another attempt to deviate from the conventional form of Japanese urban life, this time by trying to construct an inner oasis in which the occupants can shelter from the chaotic outside world. In doing so the architects have produced their own plastic concrete style.

Decorative and Innovative use of Concrete

[Left] *Takasaki Masaharu; Zero Cosmology House, Kagoshima, Kyushui, Japan (photo: Takasaki architects)*

[Below] *Zero Cosmology House*

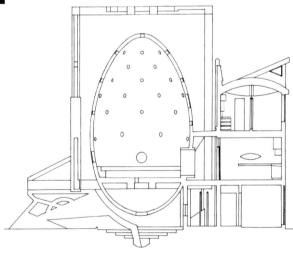

The living room on the first floor opens directly onto an enclosed courtyard. Pavers were cast by filling balloons with plastic concrete that hardened to form these hexagonal shapes. Ushida and Findlay believe that 'home is the basis for your life and it should be a very special place'. This has been achieved.

The house is constructed from a system of vertical trusses that support wire mesh that can be formed into any shape. The two layers of mesh are then pumped full of concrete. This creates a double skin that insulates the house from the heat in summer and the cold in winter.

Some of the furniture is built in, such as the seating in the living room, and the stairs to the first floor living room and rooftop patio are concrete. It is interesting to note that both this and the Zero Cosmology House have an inner ovoid-shaped room, as seen in the diagram.

5.8 Santiago Calatrava

Santiago Calatrava is perhaps the most innovative architect in current practice and of the last 20 years. He was born in 1951 in Valencia, Spain, and studied at the Architecture School and Arts and Crafts School there. After graduating he went on to study civil engineering at the Swiss Federal Institute of Technology in Zurich.

In 1981 he completed his doctorial thesis on 'The Foldability of Space Frames', and then set up his architecture and civil engineering practice. He is now an increasingly well-known and popular architect, creating striking visual architecture that, with his background in engineering, is full of movement and harmony, and with a flamboyance that might be considered by some as ostentatious.

5.8.1 Lucerne Station Hall, 1983–89

Here an existing building was in need of a lightweight glazed canopy-portico structure to extend the station along the length of the original facade. Sixteen precast concrete columns 14 metres high were placed along the front to support the lightweight roof. They are braced by steel tension columns, and an additional roof feature, rather like a wing, is held by the reverse angle of the top section of the columns. By combining alternating slopes to the columns the structure has more life than just vertical supporting columns, and the form balances well with the steel tension columns. (See http://en.Structurae.de/Structures/data/Photos.cfm?id=s0003183)

5.8.2 Swissbau Pavillion, Basel, 1988

The Swiss Association of Precast Concrete Manufacturers commissioned Calatrava to design a structure that would demonstrate modern concrete casting technology. The theme he chose was concrete in motion and he returned the challenge to them by designing a structure that called upon high-strength concrete, moulded in linoleum-lined, compound curved precision formwork.

Conceptually, the pavilion shows how static forms can be related to nature by changing their own shape through coordinated distinct movement.

Fourteen units, 7.8 metres long, tapering from 52 to 10 cm in width as well as curving in length and changing in section along the length, were cast, each weighing 1.2 tonnes. The individual units, which resemble fingers, balance on pivots that cantilever out 1.83 metres from the back support wall. Each element can be moved via a crack that connects each knuckle to a row of elliptical pins mounted on rotating disks.

The structure resembles a waving hand and is a tribute to the design and the skill of the precasters which allowed concrete to move.

5.8.3 Lyons Airport Station, 1989–94

The station is the terminus for TGV, high-speed trains, connecting the airport to Lyons some 30 kilometres away. The structure is an exciting functional use of concrete, a three-dimensional experiment in spanning, enclosure and movement.

The site was not confined and the complex comprises a main station building with platforms and a passageway connection to the airport. Entering from the upper deck, a concrete V-shaped abutment joins the ends of four steel arches that thrust towards the traveller. Above the tracks a web-like arrangement of intersecting and diagonal arches span the platform.

The structure was cast *in situ*, including the recesses for light fittings. The concrete has a light colour through the use of local white sand.

[Above and following page] *Santiago Calatrava; Lyons Airport Station, France, 1989–94 (photos: Tom Godber)*

Concrete Applications from Around the World

5.8.4 Shadow Machine at New York, 1992

A private client commissioned this work to coincide with an exhibition of Calatrava's work held at the Museum of Modern Art in New York. The Shadow Machine is a continuation of the kinetic art sculpture that was first demonstrated in the Swissbau Pavillion of 1988. This time the precast structure had 12 'fingers' that were both longer and lighter than the previous incarnation, each being 8 metres long and weighing 0.6 tonnes.

The individual units pivot on supports that are held in front of a 30 tonne base that also supports the chain drive mechanism producing the staggered synchronised motion of the Shadow Machine. As a form of sculpture the device is simple, but the principal intention was to show how a hard and heavy material can be designed to appear light and graceful, mimicking the movement of a bird or the waving hand of a ballet dancer. The curved precast elements are a form that appears later in mobile roof designs by Calatrava.

After a period at the Abby Aldrich Rockefeller Sculpture Garden in New York, the assembly was moved to Venice.

5.8.5 Auditorium in Tenerife, Santa Cruz de Tenerife, Canary Islands, Spain, 2003

This assembly of forms was commissioned in 1991 and completed 12 years later. The auditorium seats 1800 and the chamber music hall 400.

Curved white concrete shells, free-standing forms, are used for the outer walls, and the over-sailing roof resembles a wave of water from the nearby sea breaking over the top of the building. The main roof wing rises to 60 metres, supported from three concrete castings, before curving over and then thinning and converging to a single point some 98 metres from the base of the roof arc.

Concrete is used for the cluster of shells that form the main elements of the building, as well as for the other less extrovert elements of the assembly.

Calatrava takes and makes concrete into new and exciting forms that suit the material and must surely change the misconception of it as a grey and boring construction medium. It is the use of the material in these exceptional forms, as practised by Calatrava and his peers, that will help convince others of the intrinsic potential of concrete in construction.

[Above] *Santiago Calatrava; Auditorium in Tenerife, Santa Cruz de Tenerife, 2003 (photo: © Auditorio de Tenerife)*

[Opposite] *Louis Kahn; Margaret Esherick House, Philadelphia, Pennsylvania, USA, 1959–62 (photo: Amber N. Wiley)*

230

Decorative and Innovative use of Concrete

Louis Kahn; Details from the library at Phillips Exeter Academy, New Hampshire, USA, 1965–72

[Left] (photo: Yousif J. Al-Saleem)

[Below] (photo: © D. Hughto 2010)

[Below left] (photo: Yan Da)

5.9 Easton and Robertson

Behind a conventional brickwork front to the Royal Horticultural Society New Halls in London lies a dramatic interior. The roof is supported on a series of concrete arches with tiered clerestories of windows. The basic idea was apparently copied from a timber exhibition building in Gothenburg in 1923.

The building may be familiar to viewers of BBC Television as it was used in a publicity film in which acrobats were suspended from the roof on a series of ropes. The form seems ideal for concrete, acting as both columns and roof trusses in a curved transition, easily adapted to precast construction for both large and small buildings with potential for use in low cost housing.

5.10 Louis Kahn

If Louis Kahn (1901–74) had not lived beyond 1950 he is likely to have been little more than a footnote in the history of American architecture. His key contributions came following the Second World War, when he responded to the interest in a new form of 'monumental architecture', and found concrete and brick to be suitable materials to express that form. He felt that monumentality was 'a spiritual quality inherent in a structure which conveys a feeling of its eternity', and built with that in mind. Mass and geometrical form combined with penetrating light. He also had an interest in experimental structures. Louis Kahn was not prolific in his output, preferring to spend time giving great attention to detail in his creations. At the time of his death there were more prestigious commissions in his office than ever.

The Margaret Esherick House in Philadelphia was designed for a single woman and a servant. It was planned on a 3 feet 6 inch module (107 cm). The main area is 9–10 modules with a central staircase, one module wide. Not much is written about this house and the intended owner/occupier apparently died moving into it. It is not known if that was before or after inspecting the completed building. I leave that for the reader to ponder.

The walls appear to have been rendered and the three levels of render are less successful than the normal exposed raw finish preferred by Kahn.

The National Assembly Building in Dhaka, Bangladesh is another story. Kahn had been third on the list of architects after Le Corbusier and Alvar Aalto, but this turned out to be his greatest and grandest project, and was still occupying him at the time of his sudden death in 1974. The commission was active over the last 12 years of his life. The outcome of the project was that the centre of government was moved from Karachi to create a new capital, Islamabad.

Perhaps the main decision was to locate the Assembly Building separately from the rest, raise it on a platform and cast it in raw, plain concrete. The other buildings were to be in brick and at a much lower level. Casting such a structure in concrete represented a challenge in terms of organisational and technical aspects. Accordingly, Kahn designed the walling so as to accommodate a slower rate of build. A typical rate of wall height gain was determined as 5 feet (1.5 metres) a day, so the ceiling height was set at 10 feet (3 metres) and a linear marble string course feature was included in the walls at the daily pour height of 5 feet. Alternate string courses had a drip to break rainfall down the walls and reduce mould growth.

After numerous design rejections the roof to the main chamber was formed all in concrete, as an umbrella stretched between parabolic ribs. The building was not occupied until 1982, eight years after Kahn's death.

The Library at Phillips Exeter Academy in New Hampshire came approximately midway between the other two examples presented. The outer walls are clad in brickwork but the inside has a central stairway in concrete, with concrete walls penetrated by massive circular openings which reveal three floors of the library as well as the concrete beams and seating (page 232). The whole is a kind of cloister with brightly lit carrels, semi-lit floors and a mellow-lit central hall. Kahn said: 'I made the inner depth of the building like a concrete doughnut, where the books are stored away from the light'.

It is interesting to note the circular openings of the Library in repeated on the outer walling of the National Assembly Building.

5.11 Tadao Ando

It is unlikely anyone in the world of architecture is unaware of the work of Tadao Ando. His work in concrete captures spaces in a new and distinctive way, whether in churches, private homes or even shopping complexes. He is very much a follower of the work of Louis Kahn, developing that influence into his own style and interpretation of the use of concrete. Of his philosophy Ando has said:

> In all my works, light is an important controlling factor. I create enclosed spaces mainly by means of thick concrete walls. The primary reason is to create a place for the individual, a zone for oneself within society. When the external factors of a city's environment require the wall to be without openings, the interior must be especially full and satisfying.

On the subject of walls Ando has the following to say:

> Walls manifest a power that borders on the violent. They have the power to divide space, transfigure place, and create new domains. Walls are the most basic elements of architecture, but they can also be the most enriching.

Via his appreciation of the work of Le Corbusier, Ando became intrigued by the dynamism that concrete can offer. It has now become his favourite material. His use of the medium is referred to as being 'as-smooth-as-silk'. He believes (although others might not agree) that concrete mix design is not the prime factor in determining the quality of a construction. It is the formwork into which the concrete is placed.

Such a belief may have its origins, in some measure, to seven years of his teenage life spent learning the skills of carpentry from a neighbour. This led to his acceptance and expectation that wooden shuttering in Japan will not allow a drop of grout to escape through the joints, thus preventing defective surface features spoiling the form.

Tadao Ando has a reputation for his smooth-as-silk concrete to be left as struck from the mould and never painted, plastered or adorned in any way other than by the holes used for through bolts, which must be at precise uniform spacing. These aspects define his use of concrete and have become his trademark.

Two of his most highly acclaimed buildings are the Church on the Water and the Church of Light.

The Church on the Water faces an artificial lake in which a crucifix has been erected and rises up above the surface. From within the church the congregation faces the lake through a fully glazed wall that can be rolled away completely to one side. The outside is then part of the inside and the building is in harmony with nature.

> By placing a cross in a body of flowing water I wanted to express the idea of God as existing in one's heart and mind. I also wanted to create a space where one can sit and meditate.

The Church of Light is little more than a concrete box with glazed slits that intersect behind the altar to form a crucifix of light in the wall. This is the only light entering an otherwise dark interior.

The Pulitzer Foundation (http://www.pulitzerarts.org/architecture-commissioned-art/) was designed to hold the Pulitzer Collection. It is a simple assembly designed to provide the right proportions between windows, openings and walls. It consists of two separate rectangular halls, with the taller and longer one as the main gallery and the lower one housing the entrance, offices, etc. The roof is covered with a garden and the length difference is accommodated by an overhanging roof entrance.

Between the two halls is a pool that reflects the walls and creates an impression of greater length. All the concrete surfaces are in the Ando form of plain, struck concrete with regular through bolt hole spacing.

Awaji Yumebutai (http://www.yumebutai.org/english/index.html) is Tadao Ando's most sizable project. The land, in a seaside location, was first used to provide landfill for the Kansai International Airport. A total area of 215,000 m^2 has been transformed with walkways, steps, gardens and fountains, and also contains a hotel and conference centre.

Concrete has been the fundamental construction medium, and it has been tailored to the surrounding topography. It is very much a play on the use of artefacts on a desecrated coastline but the form and colour of the concrete is in tune with the surroundings, providing an appearance of sculptured rock outcrops. Of particular note is the use of scallop shells from a canning factory which were placed by hand in concrete to form the base of water features over the site.

Chicago House is in the classic Ando form of raw rectangular concrete profiles interlaid in a watery horizon.

Decorative and Innovative use of Concrete

Tadao Ando; Church of Light, Osaka, Japan (photo: Pan-O)

Concrete Applications from Around the World

This is a family house near Lincoln Park and consists of two units joined by a long narrow living room behind an inclined staircase. The smaller unit is the reception area, used for receiving guests who can from there admire the pool and terrace. It is formed from plain raw concrete, seen as beautiful by those who admire the work of Ando – the forms, locations and interplay of light and shadow. Ando presents concrete as a bland, stone-like medium, yet it is hard to see how the effects he achieves with it could be produced using any other material.

5.12 Zaha Hadid

Landscape Formation One, built entirely in concrete and glass, is also difficult to conceive of in any medium other than concrete. Although this applies to much of the work of Louis Kahn and Tadao Ando, it is even more pertinent in this building by Zaha Hadid, as the flowing lines depend on the ability of concrete to take up the uninhibited curves more than the rectangular forms favoured by Kahn and Ando.

At first the form might appear to be alien to the surroundings, like a serpent rising out of the ground, but the three distinct concrete strands form an enclosure that has been shown to fulfil its purpose. Within the building are a restaurant, offices and an exhibition area. It is set into the ground to benefit from the surrounding thermal mass, thereby reducing diurnal temperature fluctuations.

Decorative and Innovative use of Concrete

[Below and Opposite] *Zaha Hadid; Landscape Formation One, Weil an Rhein, Germany, 1999 (photo: Wojtek Gurak – www.bywojtek.net)*

Concrete Applications from Around the World

[Above and right] *Zaha Hadid; Landscape Formation One, Weil an Rhein, Germany, 1999 (photo: Wojtek Gurak – www.bywojtek.net)*

Decorative and Innovative use of Concrete

A further selection of works in concrete from around the world

5.13 Unique Hotel, São Paulo, Brazil

Ruy Ohtake, with Joao Armentano, has designed a hotel in Sao Paulo that is an amalgam of novel uses of materials and construction ideas. Post-tensioning has been used for all the floor slabs in the building. A lobby wall has dark grey on the external face and red on the inner one. This was achieved by spraying black gunite onto the outside wall shutter and then erecting the inner face shutter and pouring red concrete to form a monolithic wall with a red inner face.

Weathered copper is used to clad the facade that spans the inverted arch form of the building, pierced with circular windows. The underside is faced in wood. Post-Modernism

[Above] Unique Hotel, Brazil. The boat-shaped structure comprises of exposed concrete on the underside and exposed decorative and structural walls. (photo: Robbie Robinson)

[Below] Weathered copper clad façade. (photo: Robbie Robinson)

is expressed in the thinness of the exposed concrete vertical end walls – so thin as to appear incapable of supporting the building. The boat shape is distinctive and provides both a flat roof garden and less obstruction to the site at ground level.

5.14 Ward House, New York, USA

The William E. Ward house in Port Chester, New York, constructed between 1871 and 1876, is one of the most noteworthy achievements in the history of the use of concrete. This was the first reinforced concrete building in the USA. Both internal and external walls were cast *in situ* in concrete, as were also the internal cornice details and the castle-like tower.

[Above] *Mercer's Fonthill (photo: Michael Kendrick)*

[Opposite] *William E. Ward; Ward House, Port Chester, New York, 1871–76 (photo: Daniel Case)*

The floor beams, the floor and the roof were also made in reinforced concrete. The reinforcement used for this prototype was lightweight iron beams and rods. The building is still standing, a testament both to the durability of reinforced concrete and to the skill of those who carried out the work.

5.15 Mercer's Fonthill, Pennsylvania, USA

Another early and lasting example of reinforced concrete is found at Doylestown, Pennsylvania, USA. Here Henry Chapman Mercer built three large reinforced concrete buildings that resemble fairytale, almost Disney-like, structures, although he was not an engineer but an archaeologist, historian and ceramist – in fact, practically everything but the type of person one would expect to launch into such a venture.

The work began in 1908 with Fonthill. This building has 40 rooms and is constructed with thick concrete walls that were cast hollow. Other elements were also cast in concrete – beams, columns and a vaulted ceiling.

Mercer went a stage further and even cast some precast window frames using wire and iron bar reinforcement from the local scrap yard. He therefore pre-empted the current recycling trend by about 100 years.

5.16 National Academy of Construction, Hyderabad, India

The frame of this building is a series of pairs of inverted reinforced concrete triangles or 'W' shapes in five rows, each 10 metres apart. Within the building are exhibition areas used to display construction materials, as well as an auditorium and a library.

The central two-storey structure, with a floor area of c.8000 m², is raised up on stilts. The upper levels are reached via two inclined walkways. The whole is an interesting yet practical form, with windows shielded from direct sunlight by the sloping sides dictated by the concrete supporting frame.

5.17 Marina City, Chicago, USA

At the time (1959) these 64-storey twin towers were both the tallest residences and the tallest concrete structures in

Decorative and Innovative use of Concrete

Bertrand Goldberg; Marina City, Chicago, 1959 (photo: Hao Nguyen)

the world. Bertrand Goldberg, the designer, used a concrete core to house the services and utilities for the building – a bit like a vertical street. Lightweight circular concrete slabs project up to 14 metres from the core.

The repeating use of the concrete forms was originally to reduce construction costs, but now only adds to the striking nature and timeless appeal of the towers.

It is not until the 21st floor in each tower that the 450 apartments start (the lower levels are used for vehicle parks). By the nature of the plan form of the towers, each apartment radiates out from the central core like segments of an orange. Each flat has diverging walls and a unique, unobstructed view from the window.

Also within the complex are a 16-storey office building, a theatre for 1750 people, a 700-seat auditorium, shops, restaurants, bowling alleys, a gymnasium, swimming pool, skating rink, and marina for 700 small boats, and more.

Goldberg has said of the complex:

> Our time has made us aware that forces and strains flow in patterns which have little relationship to the rectilinear concepts of the Victorian engineers. We have become aware of the almost alive quality which our structures achieve, and we seek the forms which give the most life to our structures.
>
> … Whereas we had been talking for years about the machine in architecture, as part of the old Bauhaus tradition, it had more potential than anything we could imagine. The post-and-beam suddenly became a hangover from the Victorian tradition where the machine had been an expression of the human arm at work, from left to right and down. It felt almost like primitive looking at the machine that could create a material by a process that did not exist before – produce a magic that was not there before.

(From Heyer 1966)

We can only hope and expect he was talking about concrete.

5.18 Kantor's Chair, Poland

To appreciate the size of this chair you have to look carefully underneath, where the legs touch the ground. A full size

*Tadeusz Kantor; Kantor's Chair, Hucisko, near
Krakow, Poland (photo: Dorota Cora)*

Decorative and Innovative use of Concrete

[Above] *Obata & Nervi; St. Louis Priory Chapel, St. Louis, USA, 1962 (photo: Wampa-One)*

[Right] *Baydon Water Tower, Wiltshire, UK, 1974 (photo: Author's collection)*

chair would sit below the lower cross struts. The one above is an enormous copy, 10.3 metres tall, in precast concrete, by the renowned Polish artist Tadeusz Kantor. Kantor was creator of an avant-garde theatre, and the development for this huge chair was completed before his death in 1990. Unfortunately, there are few published details on the project.

5.19 St Louis Priory Chapel, St Louis, USA

The full name of this church, which was erected in 1962, is The Church of St Mary and St Louis and the Priory Chapel. It was designed by Gyo Obata, who was born and raised in St Louis of Japanese heritage, and working for Hellmuth, Obata + Kassabaum. Pier Luigi Nervi was retained as a consultant on the project.

The church is formed by three tiers of concentric whitewashed thin wall concrete. The first two tiers, ground level and above, are a circle of 20 arches, the ground level having a larger diameter. Each arch has been cast to form a parabola that houses an individual altar.

The top tier forms a bell tower and acts as a spire supporting a cross. The form of the church evidently reflected the methods of construction and material usage of the time. There is some resemblance to the pavilion structure cast in GRC in 1977 in Stuttgart.

The arches are faced with dark insulated fibreglass reinforced polyester, and GRP window walls create a meditative translucency when viewed from inside the church.

5.20 Baydon Water Tower, Wiltshire, UK

Travelling along the M4 motorway to the west of Reading and before reaching Swindon, one cannot fail to notice a water tower just to the south of the motorway at Baydon, Wiltshire. The tower was erected in 1974 and won an award from the Concrete Society for its design. Even now, after more than 30 years of service, it is performing well in terms of appearance and resistance to weathering, especially considering its location next to the motorway.

The water tank is in the form of an exposed aggregate, truncated, inverted cone, supported on radial concrete spokes that connect the central concrete column to the smooth-finish external concrete struts. A further connection to each strut is made about halfway down the central column but this is concealed by thickening of the struts to accommodate bending.

The appeal is in both the simplicity of the design and the quality of the build, that have together provided an impressively durable structure.

5.21 Denver Botanic Gardens, Denver, USA

At first one might not consider using concrete to construct the frame of a greenhouse. However, when the structure is large and becomes known as a conservatory, the framing section increases in size and it becomes practical to consider reinforced concrete to provide such a frame.

In the Boettcher Conservatory at Denver Botanic Gardens, lightweight reinforced concrete was used to provide the frame structure in a form of intersecting ribs which produce diamond-shaped curved window panels, some of which open.

The vaulted frame was cast *in situ* and is 49 metres long, 25 metres wide and 16 metres high at the ridge. Concrete has proven a wise material choice that performs well in the humid environment of this massive structure, as well as providing a delicate, sinuous appearance.

5.22 Gatwick Airport Acoustic Wall, Gatwick, UK

This wavy acoustic barrier at Gatwick Airport in Sussex, England is made up of 344 individual units, each 1.4 metres tall by 10 metres long and 0.5 metres thick. The wall is eight units high and 43 units long. The units were precast by CV Buchen Ltd using cement from Derby, sand from Plymouth, Devon and white granite aggregate from Dumfries, Scotland.

The design used the minimum amount of material and provided structural integrity against wind and additional blast loads whilst still being transportable from the precast yard. Each of the units was fabricated using batch-casting techniques. The wavy form and high standard of production has provided Gatwick with an attractive structure to protect the local environment. Surface finish is grit-blasted, exposed aggregate.

The wavy acoustic barrier at Gatwick Airport
(photo: Anthony Hunt Associates)

5.23 Habitat, Montreal, Canada

This development – multi-family housing via stacked modular concrete units – was designed by Moshe Safdie and built as part of Expo '67. The objective was for affordable housing, each unit having private quarters, a garden and its own balcony sited on the roof of the unit below. It was originally planned to be 900 dwellings, but only 158 were built, using 354 modules, along a harbour jetty fronting. Pedestrian streets provide horizontal circulation through the complex.

The modular design would seem to be ideal for streamlined concrete construction, with its potential to cut costs. However, the complex, expensive modular forms and the problems inevitably encountered with a new build project limited the extent and viability of the scheme.

Although in its day it failed as a viable means of providing affordable housing, it has now become rather a desirable place to live and ownership has transferred to the tenants, who purchased the development in 1985.

Decorative and Innovative use of Concrete

Moshe Safdie; Habitat, Montreal, Canada, 1922 (photo: Jason Mac http://flickr.com/zjmac)

5.24 Pergola, Guadalajara, Mexico

Smooth-finish concrete has been used on this pergola in the Centro de Seguridad Publica (Public Safety Centre) in Guadalajara, Mexico. The wavy form of the thin side beams and vertical slats forms a visually interesting structure. Each column has alternating bands of textured, bush-hammered and smooth concrete. Mexico has some vibrant concrete construction activity.

5.25 Chapel of the Holy Cross, Arizona, USA

Set between two large red rock formations in Sedona, Arizona, the Catholic Chapel of the Holy Cross was conceived by Marguerite Brunswig Staude, a student of Frank Lloyd Wright, in 1932 but she was not able to bring the idea to fruition until 1956.

The concrete chapel is set into a butte and offers spectacular views over the valley 76 metres below. The giant cross forms a symbol that appears to support the roof as it protrudes from the line of the building.

The red rocks contrast with the grey concrete that springs vertically as if growing out of them. The site has become a centre for mysticism and new age prophets.

5.26 House Shail Maharaj, Durban North, South Africa

This spectacular use of reinforced concrete is a welcome change to the more common angular profile seen in abundance elsewhere in this book. Planters and deep terraces project from the house cladding and combine with the overhanging roof to provide shelter from the direct sun, absorbing heat during the day and releasing it at night.

5.27 Spiral Stairway, Molecular Biology Research Building, Chicago, USA

Concrete has been used in a decorative form to construct spiral staircases in many ways. This double helix staircase in the Molecular Biology Research Building at the University of Illinois uses precast concrete units, one of the many applications of the medium through which its plasticity and aesthetic properties can be used to create distinctive and practical design solutions. (www.uic.edu/orgs/mbrb/staircase.htm)

5.28 Fountain of Time, Chicago, USA

Until this *Fountain of Time* sculpture in 1922 it had been customary to cast groupings as individual figures and then to assemble them on a base, allowing each to touch at a few places. However, this massive assembly was cast as one wall 37 metres long and 5.5 metres high.

The work was a collaboration between the sculptor Lorado Taft and John Joseph Earley. As discussed in Chapter 3, Earley was well known in the USA for applying decorative finishes to concrete, and in the 1930s began to use absorbent plaster moulds to improve surface detail. He also pioneered polychrome concrete in low cost housing in the

Decorative and Innovative use of Concrete

[Above] *Lorado Taft & Joseph John Earley; Fountain of Time, Chicago, 1922*

USA, but only a few houses were built, in Maryland, before his sudden death. Earley had developed a means of exposing aggregate on the facing of concrete, which was employed on this collaborative effort. The work depicts a 'human wave' overlooked by the only solitary figure, that of Father Time.

The work is part of a concrete fountain, and this wall alone was formed within shuttering that took six months to build and when erected was filled in 25 pours. The surface has weathered well, further testament to the skill of the workforce and the enduring capability of concrete. Earley also produced precast exposed aggregate panels, sculptures and huge decorative Doric columns in shell sections around load-bearing central columns for the replica Parthenon constructed in Nashville, USA in 1925 (page 45).

5.29 Christo Redemptor, Rio de Janeiro, Brazil

Towering 700 metres above Rio de Janeiro is this concrete icon of Christ, designed by the Polish sculptor Paul Landowski working in France and completed in 1931. The statue is precast concrete and the sections were erected on top of the Corcovado Mountain. The exterior has a cladding in soapstone to add colour, and construction took five years. At the base is a small chapel. The Vatican has just pronounced the structure a religious artefact, meaning the chapel can now be used for weddings.

5.30 Black Hawk, Illinois, USA

This is another creation by the sculptor Lorado Taft, working with assistance from the young sculptor John G. Prasuhn. Some of the techniques of John Joseph Earley were employed here and fibre-reinforced moulds were used as permeable formwork. The statue of Black Hawk was built in 1910 in Oregon, Illinois, and is sited high above the Rock River.

A rope was once hung from the top so that an agile person could climb up and peer through the window in the forearm. The figure stands 20 metres high from the base to the head.

The construction was carried out in winter and required heat which was supplied from radiators placed inside the

Christo Redemptor, Rio de Janeiro,
Brazil, 1931 (photo: Edward L. Weston)

Concrete Applications from Around the World

*Lorado Taft; Black Hawk, Oregon,
Illinois, 1910 (photo: Danny Higgins)*

Decorative and Innovative use of Concrete

scaffold holding the fibre moulds in place. The last of the concrete was placed in the moulds on 30 December 1910. The moulds were left *in situ* and the hydrating structure heated for a further two days. The moulding was removed in March 1911.

The outer 60 mm surface of cement pink aggregate and sand was found to be almost defect free, probably due to the superior quality of the outer skin left against the absorbent plaster moulds. The head itself, 2 metres high, was moulded against a clay master at ground level and then the mould lifted into place over the body.

5.31 Simmons Hall, Massachusetts Institute of Technology, Boston, USA

This building is in Boston, a city with perhaps more concrete buildings than any other. The first impression is of the apparent great height, a misleading one generated by the fact there are three windows per level. The building is only nine storeys high.

Heavy concrete outer walls provide natural air conditioning due to their thermal capacity which tends to even out air temperature variations between day and night. The window jambs have coloured aluminium cladding.

Entering the building one can appreciate the full beauty of the concrete exposed finishes; retired workers apparently had to be reinstated to produce the formwork. Some say the design is inspired by a sponge, with ludicrous atrium spaces, and many dislike the appearance of the building, 'like a computer chip on edge'. Some of the rooms inside have no straight lines and depict a 'concrete womb'.

The building has won numerous design awards including the 2003 Honour Award for Architecture from the American Institute of Architects.

5.32 Yanbu Cement Works and Village, Saudi Arabia

The cement works office complex and entrance is a play on the use of ferro-cement. This material has allowed artistic freedom and avoided expensive mouldings, keeping costs down. As described previously, ferro-cement is essentially layers of wire mesh supported on reinforcement and

[Above and left] *Steven Holl Architects; Simmons Hall, Massachusetts Institute of Technology, Boston; USA (photo: Scott Norsworthy)*

overlaid with mortar. It can therefore be tailored to any profile.

The village is on a gentle slope by the coast, and the design aimed to avoid a barracks settlement and create a real village. Every house has its own veranda oriented towards the sea. Streets are winding, providing a more interesting and natural layout.

Some of the houses have a steel frame with Siporex blockwork walling rendered with Vetonit whilst others are reinforced concrete columns and beams and some precast concrete panels. All walkways and roads are brushed concrete.

The objective was to create a friendly environment in the rather isolated location (80 kilometres from the nearest city) and to provide innovative and interesting designs, mainly using concrete or cementitious products. Although a limited number of house designs were used, variety has been achieved by changes in cladding, windows and doors, using a 'mix and match' approach to avoid regimentation.

5.33 Private Swimming Pool, Oxford, UK

A private swimming pool may not sound the sort of building that would be worthy of inclusion in a collection of examples of concrete excellence. However, this example is somewhat exceptional. Great thought has been put into the design and details to fully exploit the benefits of concrete in a warm humid environment. In addition, details of the design echo the features and form of the Edwardian house to which the pool is attached.

Decorative and Innovative use of Concrete

[This page and opposite] *Private Swimming Pool, Oxford, UK (photos: Edwin Trout, Information Services, The Concrete Society)*

White cement with natural calcined flint aggregate has been used in the concrete, with the aggregate kiln-burnt to remove any organic impurities. The concrete surface has been ground and polished to remove laitance and provide a smooth surface that glints slightly. The floor has been laid in white Sicilian marble.

Most major structural elements are outside the building or enclosed within the walls. Only one twin support column is located within the pool, by the changing room. Outside columns echo the mansion uprights.

Some of the novel design detailing includes: chamfered corners on the concrete elements to prevent chipping; ducts to contain wiring; anti-condensation electrical heating tape to override the cold bridge effect where concrete beams cross from the warm, humid interior atmosphere of the pool to ambient outside temperatures; secret drainage channels and drips to prevent staining; drainage downpipes eliminated by using self-cleaning chains for the water to trickle down. Note also how one of the roof beams has been continued to provide a support for the plant room flue pipe.

All joints and glazing mullion bars align perfectly. These combine with prominent framing constructed in elegant proportions, featuring protruding covers over fixing points and access to pre-stressing (better always to feature fixing locations than try to hide them). The whole presents a delightful exercise in the capabilities of concrete in the hands of thoughtful designers.

5.34 Terry Pawson, UK architect, London

Two examples now of the work of Terry Pawson, whose portfolio has widened into a national and international practice. St Mary's Church Garden Hall, Wimbledon is a simple building, with rectangular walls but to an unusual form that provides architectural interest to what might otherwise have been just a small church hall.

Vernon Street Offices offered more scope to reflect the influences of Tadao Ando and Louis Kahn – smooth concrete surfaces, the exterior of which was formed using rectangular precast concrete planks. Much care and concern was exercised in the appearance of this building, as I witnessed first-hand from a brief professional involvement during the cladding phase.

Decorative and Innovative use of Concrete

Terry Pawson Architects;
St Mary's Garden Hall,
Wimbledon, London
(photos: Tom Scott)

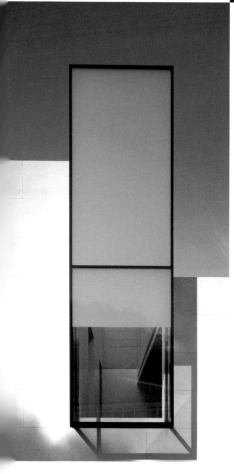

265

[Above] *Terry Pawson Architects; St Mary's Garden Hall, Wimbledon, London (photos: Tom Scott)*

Decorative and Innovative use of Concrete

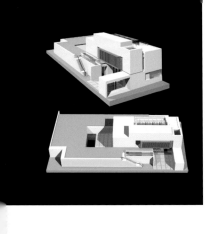

[Left, above and following page]
*Terry Pawson; New Offices – Vernon
St, Hammersmith, London, 2004–06
(photos: Terry Pawson Architects)*

Concrete Applications from Around the World

Perhaps we can categorise this style as International Modernism, bringing geometrical clarity to Hammersmith and providing vivid contrast between the smooth concrete walling and the acid-etched glazing.

It may appear that a mere concrete box has been built, but the light and airy atmosphere this type of concrete construction achieves 'bringing the outside into the interior of the offices' is remarkable. Such projects require both the involvement of a competent architect such as Terry Pawson and a reliable and consistent supply of high-quality concrete components.

31 Vernon Street has acquired three architectural awards over the last year.

5.35 UFA Cinema Complex, Dresden, Germany

This building, designed by Coop Himmelblau and constructed between 1996 and 1998, has been described as 'designed like a video-clip and seeks to do away with centralized perspective'. It would appear that concrete can be folded and twisted into any position the designer wants.

5.36 Viejo House, Santiago, Chile

This is a family house built entirely of concrete. The circular pattern of window openings is also adopted in the roof, but the roof windows are larger. Concrete provides solidity to the residence. The house walling has textural variations on the interior surfaces that have been chosen to complement the height, use and proportions. (See www.wallpaper.com/architecture/viejo-house-bymathias-Klot3/10).

This building shows that concrete can be used to construct a house in a simple, powerful but practical form and provide comfortable and luxurious living.

5.37 Arthur and Yvonne Boyd Education Centre, West Canberra, New South Wales, Australia

This building provides accommodation for students at the Art Education Centre which was built on the ranch of the late artist Arthur Boyd, 180 kilometres south of Sydney. It

[Above] Coop Himmelblau; UFA Cinema Complex, Dresden, Germany,
1996–98 (photo: Wojtek Gurak – www.bywojtek.net)

was designed and built over the period 1995–98, by Glenn Murcutt working with Wendy Lewin and Reg Lark.

Concrete floors and walls slot together under recycled roof timbers and corrugated galvanised iron roofing. The building provides low cost housing for 32 students from Australia and around the world. Projecting the concrete walls out from the front line of the building has afforded more privacy to the room units.

This building won the 1999 Sir Zelman Cowen Award for Public Buildings, the Royal Australian Institute of Architects (RAIA) most prestigious Federal Award and The Kenneth F. Brown Asia Pacific Culture and Architecture Design Award 2001.

5.38 Springwater, Seaforth, Sydney, New South Wales, Australia

This house, built between 1999 and 2002 on the Sydney Harbour foreshore, was designed by Peter Stutchbury. It is on three levels that 'finger' down the site to the harbour. It sits unobtrusively like a veil on the landscape, which provides both spectacular all-round views and shelter. The concrete frame, walling and balconies intermingle with the trees. This is a relatively low cost house, using repetitive system formwork.

5.39 Palmach Museum of History, Tel Aviv, Israel

I have tried to keep personal opinions to a minimum but admit to finding some elevations of this building by Zvi Hecker 'questionable'. The high cost of preserving the landscape apparently dictated many architectural details being left intentionally rough, so perhaps the appearance was in some way intended. There are classic examples of concrete defects spread liberally over the building, and it is not a good advert for exposed concrete. Where it has been overclad with stone the appearance has been improved. (See http://rafisegal/palmach-history-museum/)

5.40 Memorial to the Murdered Jews of Europe, Berlin, Germany

This memorial by Peter Eisenman is located in the heart of

Decorative and Innovative use of Concrete

[Above] *Arthur & Yvonne Boyd*
Education Centre, West Canberra,
New South Wales, Australia, 1995–98
(photos: Keith Saunders, 2007)

[Opposite] *Peter Stutchbury; Springwater, Seaforth, Sydney, New South Wales, Australia, 1999–2002 (photos: Peter Stutchbury architecture)*

[Above] *Peter Stutchbury; Springwater, Seaforth, Sydney, New South Wales, Australia, 1999–2002 (photos: Peter Stutchbury architecture)*

[Opposite] *Memorial to the Murdered Jews of Europe, Berlin, 2004 (photo: © Daniela Jäger/Dreamstime.com)*

Decorative and Innovative use of Concrete

Berlin close to the Brandenburg Gate and Pariser Platz (page 273). It covers two hectares and consists of 2751 blocks of grey concrete of different heights, from a few centimetres to four metres. The last units were placed in December 2004.

These precast concrete units together form a dramatic sculptural work of art with historic meaning; a cityscape graveyard. The simplicity of the arrangement translates into a powerful statement. These tombstones, for that is what they are, provide a lasting reminder of events that should not be forgotten.

5.41 Library of the Eberswalde Technical School, Eberswalde, Germany

First impressions of this library building suggest a use relating to art rather than its current function supporting forestry and

[Left and opposite] *Herzog & de Meuron; Library of the Eberswalde Technical School, Eberswalde, Germany, 1994–99; (photo: Jonas Klock)*

applied science. A ground-level connecting walkway links this new box-like building with existing ones – a building that would have little interest if it were not for the facade. The architects, Herzog & de Meuron, together with Thomas Ruff, have applied repeated images of art and photographs to some windows and all the concrete cladding panels.

The technique as applied to concrete is described in Chapter 4 (page 206). An image is produced on paper mould lining using a dot matrix layout of chemical retarder. After demoulding, the retarded surface is washed off, leaving the etched image.

5.42 Holy Rosary Catholic Church Complex, St. Amant, Louisiana, USA

In this complex, designed by Trahan Architects, 2000–04, a minimal amount of landscaping has been combined with

Trahan Architects; Holy Rosary Catholic Chruch Complex, St Amant,
Louisiana, USA, 2000–04 (photos: Trahan Architects)

the concrete architectural form, devoid of any colour, to enhance a religious function and purpose.

The client required an appearance that was neither austere nor ostentatious and had a preference for low cost materials, hence the use of unadorned raw concrete.

Architects seem to follow trends, picking out ideas that have worked well elsewhere. Here the plain concrete, shuttering marked surfaces are reminiscent of the work of Tadao Ando, though they are linked in a less regimented form. In such applications plain concrete is not used as a decorative form in a conventional way, but rather has to be seen in the context of the way the surface forms are played upon by light.

5.43 Baumschulenweg Crematorium, Berlin, Germany

Treptow Crematorium, Baumschulenweg, designed by Axel Schultes and Charlotte Frank, 1992–98, is another example of the use of Modernist style to create a temple, or in this case a crematorium complex, in white plain concrete. (See www.architectureinberlin.com/?p=66)

This building has become a centre of interest for students and enthusiasts of contemporary concrete architecture. There is clearly a Tadao Ando influence, though the finish quality of the concrete is perhaps not as high. Twenty-nine concrete columns rise, as if made of marble, to a ceiling where the capitals allow light to break through and adorn the space.

5.44 Grassi House, Lugano, Ticino Canton, Switzerland

Reinforced concrete, apparently so simple in form, has given this house clean lines and a form that fits well into the location. Concrete thickness has been reduced and the metal window frames concealed so the windows appear as interruptions in the elevations. The scale and proportions are just right to the eye. (See www.guscetti.ch/ [see Casa a Lugano])

The house was built in 2001 and is the work of Giovanni Guscetti. Guscetti has produced a subtle yet confident design that provides a strong physical appearance. The secret of Guscetti's successful use of concrete lies in the proportions, and perhaps also in the low rise. If it had been bigger, it might have been too dramatic in such a location.

5.45 Convention Centre, Barcelona, Spain

In contrast to the Grassi House, the Convention Centre in Barcelona by Carlos Rerrater and Jose Maria Cartana is big. A concrete statement such as this has to be considered in a fragmented form, especially as it is laid out in three bodies over the sloping topography. White concrete is used over the entire external envelope.

5.46 Booster Pump Station East, Amsterdam East, The Netherlands

Buildings can be fun! Well, how else can the design of a sewage pumping station be approached? There are too many drab and ugly attempts at designing functional buildings. With this one, built between 2003 and 2005, Bekkering Adams provided a concrete solution that can be considered as sculpture. The building had to contain the noise generated within, so concrete was an ideal choice. A crystal-like form is combined with blue-green and marble pigmented concrete, grit-blasted with graphics over the walls and roof and moulded in relief around the base with plays on the word 'Booster'. (See www.bekkeringadams.nl [see project in Public and Education])

This final chapter provides examples of some recent developments in concrete. Some have just seen the light of day while others have recently been put into production and use, and together they give some indication of how concrete might be used decoratively and innovatively in the future.

6.1 Ferro-cement

The first application is something of a leap back in history. A new type of cement, discussed later in 6.2, could be used to rejuvenate a thin reinforced concrete. *Orihune*, the world's first folded concrete canoe, was an exercise in producing what must be the thinnest form of reinforced concrete possible – 4 mm thick.

The original objective was to construct a seaworthy canoe in three weeks. The idea was developed of producing a foldable sheet that when folded, and the joints sealed, could perform a structural role. The advantages of such a technique are:

- light construction
- simple flat formwork
- excellent control over thickness
- high-quality finish of the form face
- no special tools or equipment needed
- quick to produce in one operation.

Two sheets of 0.8 mm diameter mesh were laid over the mould and screeded with mortar. Joint locations were formed by pre-locating foam strips on the form and clamping the mesh down with a timber strip over the foam strips to compress the foam into the mesh. After curing, the sheet was removed from the form board. The hull was then folded at the exposed mesh joints and the joints filled against masking tape applied to the inner face.

This simple means of construction could be updated by incorporating alternative reinforcements such as carbon fibre mesh or glass fibre, or combinations of reinforcement. The cementitious matrix could also be updated, to incorporate much lower cement content and include microsilica and polymer additions.

The technique is constrained only by the limits of imagination. The inventor, Robert J. Wheen, considered it could be used for constructing yachts, barges and pontoons, as well as, on land, box culverts, barbecues, incinerators and garden furniture. Other possibilities include permanent formwork for columns or façade treatments, and it could even be formed into box girder segments.

6.2 Ultra-high performance concretes

In 1981, Birchall, Howard and Kendall, of ICI Mond Division, published a development they called 'defect-free concrete'. They showed that it was possible to produce concretes with phenomenally high strength using cement, microsilica, polymers and much reduced water content. Key to the success of these concretes was the removal of macro-defects in the structure of the hydrated cementitious paste enabling the cement to be formed into helical springs that deflected and maintained load.

Such concretes are now available under the collective name of ultra-high performance fibre reinforced concrete (UHPFRC). Most manufacturers have their own brand name, used to distinguish their product from others. The common aspects are a high steel fibre content, use of selected sands, and addition of fillers, in particular microsilica and polymer, to enable ultra low water content.

Compressive strengths are normally in the range 150–300 N/mm^2, and direct tension strengths 10–15 N/mm^2. Cement contents are at the level circa 900 kg/m^3 and modulus 45–55 kN/mm^2. Shrinkage is high due to the high cement inclusion, at 500 microstrain. However, any subsequent shrinkage can be eliminated by heat treating the material.

Uses so far have been divided between prestressed and passive steel fibre reinforcement applications and include:

- beams
- columns
- footbridges
- façade panels
- street furniture
- canopy roofs
- joints (the material can be used as a jointing between elements)
- drain covers
- different forms of staircase.

6.2.1 Lafarge Ductal®

Lafarge has developed the macro-defect free concept with Ductal®, their own formulation of cement, silica fume, mineral nanofibres, metallic or polyvinyl acetate fibres, superplasticisers, water and sand.

The Martel Tree

The Martel Tree was originally designed in the 1930s by brothers Jean and Noel Martel and was at that time to have been made in steel. In 1999, on behalf of Boluogne Billancourt City, France, it was redesigned by architect A. Spielman for construction in Ductal®. Engineered by Bouygues, a partner of Lafarge in the Ductal® development, it is 8.5 metres tall with some pieces 6 cm thick.

Chairs for the future in New York

In 2005 the International Contemporary Furniture Fair in New York exhibited a new line of ultra thin and light chairs made in Ductal®. They were made by Formglas, one of the North American leaders in manufacturing prestressed products.

The chairs were designed by the architect Omer Arbel, who made use of the thinness of Ductal® and its ability to be coloured throughout. The units are c.5 mm thick.

Shawnessy Light Rail Transit Station, Calgary, Alberta, Canada

The CPV Group Architects and Engineers used Ductal® in their design for a new light rail transit station at Calgary, Alberta. The station includes 24 structures precast in Ductal®, including canopies, struts, columns, beams and gutters.

Ductal 8.0 chair (photo: Shannon Loewen)

Perforated panels

The strength of Ductal® has enabled the erection of perforated mesh panels as light screens in a nursery school in Ornans, a small village in the Doubs region south of Besançon. Twenty-three screens 1.7 metres × 3.6 metres, and some 1.7 metres × 4.6 metres, were fixed to the walls by threaded studding. The leaf motifs were produced by casting against leaves placed in the base of the moulds.

6.2.2 CRC Aalborg Portland: Spiral staircase, Tuborg Nord

A version of UHPFRC was developed by the Cement and Concrete Laboratory of Aalborg Portland in 1986. It combines a large proportion of closely packed steel reinforcement with microsilica and water/cement ratios of typically 0.16, giving composite high strengths of between 150 and 400 N/mm^2. The matrix is very ductile due to the high steel fibre content, preventing large cracks under service conditions.

There have been various cantilevered applications of the product, such as balconies for a development near Copenhagen. The superior durability of the matrix has been found to be due to a low total porosity of 1.5% for the CRC-Matrix, of which only 14% of this is due to capillary porosity. This means there are no freezing and thawing problems and carbonation rates are extremely low.

One stunning application is a spiral staircase erected at Tuborg Nord's Company House at Tuborg Boulevard, Hellerup, a suburb of Copenhagen (see www.crc-tech.com). This staircase has no central column but instead a vertical spiral beam, 1.5 metres high and 15 cm thick and cast to a 1.5 metre radius, spiralling one turn between floors. From this the treads radiate 2.25 metres. Each tread tapers from 10 cm at the root to 3 cm at the outer tip.

The assembly was cast in sections of six treads per section that provide one quarter of each flight between floors. The units were bonded together with a development of CRC called JointCast, which has higher cement and microsilica content and 6% fibre inclusion (475 kg/m^3). This acts as strong glue such that rebar requires only eight diameters to provide a full lap bond. The material cost is high, at about 10 times conventional concrete.

6.3 Italcementi Bianco TX Millennium cement (self-cleaning concrete)

Self-cleaning concrete: This is the claim of Italcementi, who have developed a titanium dioxide white cement which apparently retains its appearance over time in a way as yet unparalleled by other products.

Bianco TX Millennium is the result of laboratory research that has produced a concrete surface which oxidises organic air pollutants by photocatalysis to carbon dioxide. The pollutants therefore lack a sub-layer and do not adhere to the concrete.

Photocatalysis is receiving attention because of its ability to purify air and water and its antibacterial properties. The first applications were in Japan, and included self-cleaning construction items.

One recent application of the cement is the Dives in Misericordia Church in Rome.

6.4 LitraCon™ (light-transmitting concrete)

Hungarian architect Áron Losonczi has developed a concrete that transmits light. Glass fibres, which transmit the light, are cast into the concrete in alternating layers with the concrete. Fibres of various diameters are used, ranging from 2 μm to 2 mm. The diameter of the fibres controls the illumination effect obtained.

Light can be transmitted up to 20 metres. It is envisaged that translucent concrete walls could be produced, illuminated from the floor below. A design for a police college in Kuwait City has been proposed in which the walls provide thermal capacity, maintaining protection from the desert heat, while also letting light into the building.

6.5 Precast concrete

It is likely that in future more concrete will be provided as precast, since the quality of construction can be controlled more easily in a factory environment than on site. The precast units can be inspected and tested or sampled before delivery, so that only quality components are supplied.

One application, for walling, abutments, landscaping and retaining walls, utilises large blocks that fit together as

[Above] *Litracon™ light transmitting concrete (photo: Litracon www.litracon.hu)*

[Opposite] *Big Block system (Redi-Rock)*

Recent and Future Developments

giant building blocks, *c.*700 kg in weight for free-standing walling and 1050 kg for retaining walling.

6.6 Aircrete blocks and thin joint masonry

6.6.1 Aircrete

Aircrete is the current trade name for autoclaved aerated concrete (AAC). It is not always appreciated that foamed lightweight systems have been employed since Roman times. The first aerated concrete was patented by a Czechoslovakian, E. Hoffman, in 1889. This used oxygen as the foaming agent, or gas formed by combining hydrochloric acid and limestone.

Around 1914 Aylsworth & Dyer in the USA used hydrogen gas, produced by mixing aluminium powder with calcium hydroxide, as the aeration agent. In 1917 a Dutch patent claimed yeast as an aerating agent and later patents have cited use of zinc powder with hydrating cement, hydrogen peroxide, or sodium or calcium hypochlorite to produce stable foamed concrete.

Work by Grosahe in Berlin has found that aluminium flake powder, with Portland cement and sometimes an alkali addition, is the most reliable generator of consistently sized foam bubbles. However, the structure formed is of low strength, friable and certainly not suitable for structural use.

A major development occurred in Sweden in 1923, when Dr Johan Axel Eriksson, an architectural science lecturer, autoclaved some aerated concrete samples of burnt shale limestone, water and aluminium powder to speed up the curing process. This overnight autoclaving resulted in a significant increase in strength due to a stronger crystalline structure formed in the heat and pressure in the steam autoclave. The improvement in properties was the result of silica combining with lime to form a stable calcium silicate hydrate crystal structure, resembling the volcanic rock Tobermorite, found near the village of that name on the Scottish island of Mull.

In 1924 Eriksson patented his product, known locally as 'poren betong', and in 1929 a plant was opened by the building product manufacturer Karl August Carlen, in the town of Yxhult in Sweden. The product became known as Y-Tong®, formed from an abbreviation of the town where production took place and the Swedish word for concrete, 'betong'. Production continued until the late 1960s. Others developed the system and production of AAC now occurs worldwide.

In the UK, uptake and application of AAC has been, perhaps surprisingly, somewhat slow. Roofing panels, cladding and other materials have been produced with varying degrees of success. In roof units made using the autoclaved product, reinforcement corrosion was found to develop rapidly in service, mainly due to the reduced alkalinity.

It is worth being aware that the combination of accurate AAC masonry blocks and the new thin joint mortar has many innovative and significant advantages over conventional block masonry. These include:

- Lightness, allowing easier handling and larger blocks to be used on site without assisted handling equipment.
- Quicker build rates: Independently observed speed trials have shown that Aircrete thin joint walling can be put up twice as fast as when using aggregate blocks and traditional mortar.
- Better thermal insulation: The thermal insulation of Aircrete is up to eight times higher than dense concrete blockwork. Thin joint mortar, with joints 2–3 mm against 10 mm for conventional mortar, further improves the thermal performance.
- Ease of cutting and fixing; this speeds up installation of services and finishing.

6.6.2 Building with thin joint Aircrete

Aircrete blocks are available in a range of densities, sizes and thicknesses. Manufacturers produce scratch marks on the faces of blocks to allow identification of the various types. There are now foundation blocks available for use below damp-proof membranes – up to 355 mm wide – as an alternative to cavity wall foundation with concrete infill, engineering bricks or heavy aggregate blocks. Another advantage is a tongue and groove detail that removes the need for mortar between the perp-ends.

The structure of Aircrete enables cutting on site using coarse tooth handsaws, though some builders use hand cutting jigs and others bandsaws to ensure precise cuts.

Aircrete block manufacturers have developed dry pack mortars that are blended with water on site and applied over runs of blocks in a 2–3 mm thick layer. Application of the mortar is speeded up using either special sledges that extrude a ridged layer of mortar, or hand scoops colour coded for identification to suit particular block widths.

Speed of laying Aircrete has now been increased further by using pumped mortar applied via special sledges. Bed-joint reinforcement is used above and below openings to distribute stress and prevent cracking. The reinforcement can be thin wire or mesh, enabling it to be contained within the thin mortar joint.

Thin layer mortar is formulated so that it sets and stiffens within about 60 minutes, but only about 10 minutes *in situ* between blocks. This enables thin joint masonry walls to continue rising when conventional block masonry, after a few courses, would require to be left overnight.

When thin joint masonry is built on conventional blockwork, concrete trench or raft foundations, the first course must be laid on conventional mortar to accommodate the variation needed to align the top of the blocks. From then on thin layer mortar allows Aircrete blocks to be laid rapidly, such that a complete inner leaf of a three bedroom house can be completed in 16 hours using a standard team of two block layers and one labourer. This method of rapidly constructing the inner leaf of a house is used as one of the Demonstration Housing Forum Projects, on a site in conjunction with Gleeson Homes (North East).

Due to the rapid building rate of thin joint masonry, some builders in the UK have used thick solid Aircrete blockwork to construct housing. The resulting structure achieved a wall U-value of 0.3 W/m²K and the maximum SAP rating of 100. Time taken to construct walling and seal the roof was reduced from five weeks to five days. Aircrete is an intrinsically more efficient insulation material than solid aggregate blocks, accounting for the popularity of solid wall Aircrete construction in Scandinavia and northern Europe.

6.6.3 Finishes on Aircrete

Internally, Aircrete can be plastered, dry lined, tiled or (if the paint-grade smooth blocks are used) simply painted.

Externally, in the UK at least, it is customary to cover up blockwork with some kind of applied finish. This can be brickwork, render and tile-hanging over felt and battens.

However, external coating systems that can be applied directly onto Aircrete masonry walling, without the need for prior rendering, are now available. Some of these are reinforced and others are cement free and can accommodate any cracking without transmitting the cracks across the surface.

The Wonderwall® system, developed by Hanson Building Products, comprises an insulating panel pre-bonded to a vacuum-formed brickwork coordinating carrier sheet. This composite panel is first fixed directly to the Aircrete wall using proprietary fixings. Brick slips are then bonded to the carrier sheet using a special adhesive and the joints finished off with a specially developed pointing mortar.

6.6.4 Cautionary notes

When using Aircrete it is important to install vertical movement joints at least every 6 metres, or less on external walling, accompanied by debonded wall ties across every other vertical joint. In addition, the first joint down a wall should be within 3 metres of the start or edge of a wall.

All openings, such as for windows, should be reinforced with bed-joint reinforcement, 1.5 mm thick, laid above and below the opening in two bed-joints extending at least 600 mm beyond the opening. Optimum performance is achieved by installing the bed-joint reinforcement continuously around the building at these levels.

Special wall ties are required to allow installation within the narrow bed-joints. Helical ties that can be hammered into Aircrete are available. Additional ties will be needed around openings and at movement joints where cavity construction is employed.

Any application of Aircrete should comply with the published recommendations of the Aircrete Products Association and the recommendations of the Aircrete producer.

6.7 Insulated concrete formwork

Insulated concrete formwork (ICF) is a system of providing supporting formwork for concrete to build loadbearing and non-loadbearing internal and external walls for buildings. Once the concrete has set and becomes loadbearing, the formwork takes up the role of insulating the structure and providing a surface for the application of finishes.

Insulated concrete formwork is essentially a twin-wall of expanded polystyrene (EPS) that is available in either block or panel forms. The concept is not new and has been in use on the continent and North America for many years.

6.7.1 Block systems

An example of the Beco Wallform block system of ICF is shown. The two wall panels are held apart by a rigid cross brace moulded into the block form. Various sizes are available, with widths ranging from 250 to 438 mm and lengths from 1000 to 1250 mm. The top and bottom of each block has tongues and grooves, respectively, moulded into the form to allow the units to link together. This prevents the units from blowing away as they are erected on site.

The vertical faces of the blocks have dovetail grooves to provide a mechanical link for any applied finish. End of runs are blanked off with closure panels. Angled corners, curved wall units and various block thicknesses are available.

6.7.2 Panel systems

Another form of ICF is the panel system, where larger parallel panels of EPS are braced apart to receive the internal pour of concrete. The Quad-Lock panel system, uses only four components: panels, ties, corner brackets and track. This American system has Regular panels 48 inches long × 12 inches high and 2.25 inches thick and Plus panels 4.25 inches thick.

The options are, therefore, two Regular or Plus panels, or one of each type, using ground track and ties to form the appropriate panel system. A further option is panels with fastening strips moulded into the panels at 12 inch centres.

The Quad-Lock system is particularly suited to curved forms. Wall thickness can vary between 210 mm o/a, with a concrete core of 96 mm using two 57 mm EPS panels, and 413 mm o/a, with combinations of 57 mm and 108 mm EPS panels giving concrete core thicknesses of 197 mm, 248 mm or 299 mm.

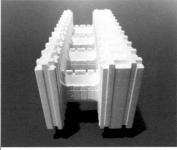

*Beco Wallform polystyrene
formwork block (photos: Beco
Products Ltd)*

6.7.3 Erecting and concreting ICF systems

All ICF systems require temporary external support whilst the
concrete is fluid, to maintain the units vertical and aligned
and also provide access for the concrete gang. Gaps in the
shuttering are filled with expanding polyurethane foam and
the concrete is pumped into shutters with the discharge
pipe within the shutter restricting concrete falls to 2 metres
maximum, and using a discharge hose of 100 mm diameter.

Concrete is normally placed in ICF systems in layers of
1 metre at a discharge rate of 3 metres/hour. Horizontal
curved walls are restricted to 1 metre/hour. Class RC25
pumpable concrete with an S2 or S3 slump or consistence is
commonly used, using 10 mm aggregate.

Reinforcement is normally needed only over three
storeys but this will of course be governed by the design.

[Above] *Placing concrete within an ICF system, Quad-lock, (England) Ltd (photo: Quad-Lock Insulating Concrete Formwork)*

[Right] *Polysteel ICF system (photo: Polysteel)*

[Below and opposite page] *Finished rendered ICF system (photo: Polysteel)*

When reinforcement is incorporated, consideration must be given to the ease with which concrete can be placed, and this will likely determine the width of the concrete core.

External finishes can be sand/cement render, acrylic render, brick or stone slips, standard masonry brick or stonework. Internally the finish is typically 12.5 mm plasterboard or a dry-lined finish with or without a plaster skim coat.

The speed of erection, the discipline the forms provide (especially to curved walling projects) and the improved insulation from the EPS mean that ICF systems offer huge potential for designers to create exciting structures in a cost-effective way.

Chapter 3

Allan, J. (2002). *Berthold Lubetkin.* Merrell, London.

American Concrete Institute (2004). *Concrete: A Pictorial Celebration.* American Concrete Institute, Michigan.

Childe, H.L. (1949). *Concrete Products and Cast Stone.* Concrete Publications Ltd, London.

Coe, P. and Reading, M. (1981). *Lubetkin and Tecton: Architecture and Social Commitment.* University of Bristol, Bristol.

Dawson, S. (1995). *Cast in Concrete: Reconstructed Concrete Stone and Precast Stone: A guide for architects.* Architectural Cladding Association, London.

Glanville, W.H. (ed.) (1939). *Modern Concrete Construction.* The Caxton Publishing Company Ltd, London.

Gray, W.S. and Childe, H.L. (1948). *Concrete Surface Finishes, Renderings and Terrazzo.* Concrete Publications Ltd, London.

Hobbs, C. (1971). *Faircrete: An Application of Fibrous Concrete. Prospects for fibre reinforced construction materials.* Proceedings of an International Building Exhibition Conference sponsored by the Building Research Station, Olympia, London, 24 November 1971.

Kelly, A. (1990). *Mrs Coade's Stone.* Self-Publishing Association.

Larkin, D. and Pfeiffer, B.B. (eds) (1999). *Frank Lloyd Wright: Master Builder.* Thames & Hudson, London.

Nervi, P.L. (1966). *Aesthetics and Technology in Building.* Translated from the Italian by R. Einaudi. Harvard University Press, Cambridge.

Powers, A. (2005). *Modern: The Modern Movement in Britain.* Merrell, London.

Ritchie, T. (1978). Roman stone and other decorative artificial stones. *Bulletin of the Association for Preservation Technology* 10: 20–34.

Stanley, C.C. (1979). *Highlights in the History of Concrete.* Cement and Concrete Association, Wexham Springs, Slough.

Stanley, C.C. and Bond, G. (1999). *Concrete through the Ages.* British Cement Association, Crowthorne, Berkshire.

BIBLIOGRAPHY

Stäubli, W. (1966). *Brasilia.* Universe Books, New York.

Welsh, J. (1995). *Modern House.* Phaidon Press, London.

Chapter 4

Aberdeen Group (1995). *Architectural and Decorative Concrete Flatwork.* The Aberdeen Group Reprint Collection, Illinois.

ACIFC (2001). *Dry Shake Finishes for Concrete Industrial Floors: An Introductory Guide prepared by a working party of the Association of Concrete Industrial Flooring Contractors.* Supplement to *Concrete* 35(10).

Anon. (1991). Federico Assler: Concrete sculptures – an artist's medium. *Concrete* 25(2): 28–29.

Anon. (1994). Concrete World: Concrete sculpture celebrates the home. *Concrete* 28(1): 4.

Ascent Publishing Ltd (2006). *Homebuilding & Renovating Magazine* July 2006. A Centaur Holdings Company.

BS EN 12878:2005: *Pigments for the colouring of building materials based on cement and/or lime – Specifications and methods of test.*

Coghill, I. (2001). Pargeting and public art. *Concrete* 35(7): 44–45.

Concrete Society Award, High Commendation (1989). Colour underfoot. *Concrete* 23(6): 31–33.

Deane, C. (2003). *Concrete Works* by Carole Vincent. *Concrete* 37(8): 8.

Gage, M. (1974). *Guide to Exposed Concrete Finishes.* Architectural Press, London.

Gage, M. and Vandenberg, M. (1975). *Hard Landscape in Concrete.* Architectural Press, London.

Gaventa, S. (2006). *Concrete Design.* Mitchell Beazley in association with Blue Circle, London.

Harris, B. (2004). *Guide to Stained Concrete Interior Floors.* The Bob Harris Decorative Concrete Collection. Decorative Concrete Institute, Inc., Temple, Georgia.

Hart, I. (2000). Photo-engraved concrete. In: Quality and special finishes for concrete. *Concrete* 34(4): 17–19.

Heyer, P. (1966). *Architects on Architecture: New Directions in America.* Walker, New York.

Ireland, J. (2005). Pebbles leading up the garden path. *Concrete* 39(7): 19–20.

Jackson, N. and Johnson, C. (2000). *Australian Architecture Now.* Thames & Hudson, London.

Kennedy, P. (2005). Monolithic concrete given organic natural essence. *Concrete* 39(7): 21–22.

King, J. (1999). Concrete luggage sculpture. In: Luxton, A., Building a rainbow. *Concrete* 33(6): 32.

Lazenby, E. (2000). Breaking the boundaries of function and art with decorative concrete. *Concrete* 34(10): 52–53.

Levitt, M. (1982). *Precast Concrete: Materials, Manufacture, Properties and Usage.* Applied Science Publishers, London and New Jersey.

Levitt, M. (1998). *Pigments for Concrete and Mortar.* Concrete Society Current Practice Sheet No. 113.

Lynsdale, C.J. and Cabrera, J.G. (1989). Coloured concrete: A state of the art review. *Concrete* 23(7): 29–34.

MacCraith, S. (2000). Concrete sculptures in Ireland. *Concrete* 34(10): 32–36.

Monks, W. (1985). *Tooled Concrete Finishes.* Appearance Matters 9. Cement and Concrete Association, Wexham Springs, Slough.

Monks, W. (1986). *Textured and Profiled Concrete Finishes.* Appearance Matters 7. Cement and Concrete Association, Wexham Springs, Slough.

Portland Cement Association (2004). *Finishing Concrete with Colour and Texture.* PCA Serial No. 2416a. Portland Cement Association, Skokie, Illinois.

Rogers, C. (2000). The Time of Day. In: MacCraith, S., Concrete sculptures in Ireland. *Concrete* 34(10): 32–36.

Stuart, I. (2000). The Group of School Children. In: MacCraith, S., Concrete sculptures in Ireland. *Concrete* 34(10): 32–36.

Undery, D. (2000). Touch me – I'm concrete. *Concrete* 34(10): 31.

Undery, D. (2001). Living with concrete. *Concrete* 35(7): 42–43.

Undery, D. (2003). The gentle art of concrete. *Concrete* 37(10): 67.

Vincent, C. (1997). Colour in three dimensions. *Concrete* 31(4): 22–23.

Vincent, C. (2001). Blue Circle Garden – Chelsea Flower Show 2001. *Concrete* 35(9): 48–50.

Vincent, C. (2002). Trials with self-compacting concrete for sculpture. *Concrete* 36(7): 36–38.

Vincent, C. (2003). *Concrete Works.* Alison Hodge, Newmill, Penzance.

Weston, R. (2008). *Materials, Form and Architecture.* Laurence King Publishing, London.

Chapter 5

Bell, J. (2006). *21st Century House.* Lawrence Bell Publishing, London.

Bennett, D. (2003). High-strength mix on a spiral of success. *Concrete Quarterly* Autumn: 10–13.

Concrete Society Award, Joint Winner (1989). First commission a winner. *Concrete* 23(6): 27–29.

Concrete Society Award, Joint Winner (1992). In: Walker, M., Eight

Awards – so far. Joint Winner 1989. *Concrete* 26(2): 32.

Jodidio, P. (ed.) (2003). *Icons – Architecture Now*. Taschen, Köln.

Jodidio, P. (2007). *Architecture Now! 4*. Taschen, Köln.

Jodidio, P. (2009). *Architecture Now!* Taschen, Köln.

Menking, W. (2003). Holl picks holes in masonry tradition. *Concrete Quarterly* Winter: 4–7.

Petterson, N. (2004). *New Minimalist Architecture*. Harper Collins Publishers, London.

Rattenbury, K., Bevan, R. and Long, K. (2006). *Architects Today*. Lawrence King Publishing, London.

Rykwert, R. (2001). *Louis Kahn*. Harry N. Abrams, New York.

Tzonis, A. (2002). *Santiago Calatrava: The Poetics of Movement*. Thames & Hudson, London.

Tzonis, A. (2007). *Santiago Calatrava: The Complete Works*. Rizzoli International Publications, New York.

Welsh, J. (1995). *Modern House*. Phaidon Press, London.

Chapter 6

Birchall, J.D., Howard, A.J. and Kendall, K. (1982). A cement spring. *Journal of Materials Science Letters* 1(3): 125–126.

Italcementi (2004). *The new white cement Bianco TX Millennium for the Dives in Misericordia Church*. Press release, 24 October, Rome.

Jaremko, D. (2003). New Shawnessy LRT Station uses 'revolutionary material'. *Alberta Construction Magazine* November/December: 105–106.

Perry, V.H. and Zakariasen, D. (2004). First use of ultra-high performance concrete for an innovative train station canopy. *Concrete Technology Today (Newsletter of the Portland Cement Association)* 25(2): 1–2.

Wheen, R.J. and Bridge, R.Q. (1981). *Orihune* – the world's first folded concrete canoe. *Concrete International* October: 32–34.

Web addresses

Aircrete Products Association: www.aircrete.co.uk

H+H Celcon: www.hhcelcon.co.uk

Lafarge Ductal®: www.Ductal-Lafarge.com

LitraCon™: www.Litracon.hu

Redi-Rock: www.redi-rock.com

Thermalite® – Hanson HeidelbergCement Group: www.heidelbergcement.com/uk/en/hanson/products/blocks/aircrete_blocks/index.htm

305

INDEX

Decorative and Innovative use of Concrete